BLANKET PARTY IN DESERT STORM

FROM BEATDOWN AND SPIRITUALLY BROKEN TO ETERNALLY BLESSED

I0520575

DARNNELL REESE & DEIDRA WILSON

FIRST EDITION NOVEMBER 2025

ISBN: 979-8-9938536-0-4

CONNECT WITH DARNNELL REESE:

INSTAGRAM: @DARNNELLS
WEB: HTTPS://REESEAUTHOR.COM
PUBLISHED BY VICTORIOUS WITH GOD
FORT WASHINGTON, MARYLAND, USA
AMAZON AUTHOR PAGE: HTTPS://AMAZON.COM/AUTHOR/DARNNELLREESE
LINKEDIN: WWW.LINKEDIN.COM/IN/DARNNELL-REESE

CONNECT WITH DEIDRA WILSON:

INSTAGRAM: @DEIDRA_ADAMS
WEB: HTTPS://WWW.DEIDRA-WILSON.COM/
LINKEDIN: HTTPS://WWW.LINKEDIN.COM/IN/DEIDRADAMS/

AUTHOR'S NOTE

THIS IS A WORK OF MEMOIR BASED ON MY EXPERIENCES DURING MY SERVICE IN THE UNITED STATES ARMY FROM 1989 TO 1992. I HAVE RECREATED EVENTS AND CONVERSATIONS FROM MY MEMORIES. SOME NAMES AND IDENTIFYING DETAILS HAVE BEEN CHANGED TO PROTECT THE PRIVACY OF INDIVIDUALS.

DEDICATION

FOR MY MOTHER, CHERYL, WHO SPOKE DEIDRA INTO EXISTENCE AND WHOSE PRAYERS BROUGHT ME HOME SAFELY FROM THE WAR.

FOR MY DAUGHTER, DEIDRA, WHO GAVE ME A REASON TO CHOOSE LIFE.

FOR MY AUNT TERESA, WHO TOOK ME IN AND GAVE ME A SAFE SPACE WHEN I HAD NOWHERE ELSE TO GO.

Prologue: Through My Eyes (Deidra Wilson)

For most of my life, I knew my mother's story in fragments. I knew it through her old camouflage jacket with her last name—Adams—still stitched across the chest. Through photographs of her squatting beside the back of a Humvee, helmet strapped tight, goggles resting against her face, fully geared for war. Through another image of her sitting on the floor, beside her cot, dog tags hanging from her neck as she polished her boots with a cheerful disposition and through her framed   Certificate of Commendation for Operation Desert Storm. These pieces of her life appeared throughout our home—occasionally displayed, but often tucked away in cupboards, buried beneath years of paperwork, or shoved deep into coat closets.

These were the artifacts of her service; the only way I really knew her war. I'm sure when I was younger, she showed them to me with pride—perhaps eager to share her history, our history. But I was too young to understand what it meant to be an Army veteran; too young to grasp the weight behind that title. Too young to understand what it truly meant for her to still be alive. What I did not know then—what I would only come to understand much later—was the hardship, the prejudice, and the isolation she experienced while serving. And that it did not come from foreign enemies, but from within her own ranks. The saddest truth I would eventually learn is that her story is not unique among women in the military.

As I grew up, I watched her fight the Department of Veterans Affairs for more than a decade, battling for recognition of the disabilities and idiopathic illnesses caused by her exposure to gases and medication during her service. Even through the eyes of a child, I could see that she was carrying something heavy. I could see the toll that endless red tape, shifting bylaws, and bureaucratic indifference were taking on her   mentally and emotionally. But I was still a child—and I wanted my mommy. I wanted her time. I wanted to see her happy. I did not want her consumed by a system that seemed designed to wear her down and chip away at the person who was my world.

It was not until I was older—college-aged—that I began to publicly acknowledge her service on Veterans Day. I'm not sure what sparked the change. Maybe it was the surge of military films at the time— *Thank You for Your Service*, *Sand Castle*, and *Megan Leavey*. But I remember feeling an unexpected irritation whenever I told someone my mom had served. I was almost always met with surprise— Really? but nothing more. No follow-up questions. No curiosity about where she was stationed, what she did, or where she was deployed. At the same time, some of those people sharing their own service, often unsolicited, seemed to expect a kind of reverence in return. That disconnect stayed with me.

Nearly ten years later, that frustration resurfaced in a conversation with my husband. He was interested in writing a   screenplay about a famous military dog involved in the takedown of a notorious terrorist and founder of al-Queda. I hadn't expressed my subtle contempt for his idea early on, so it quietly festered. Until one day I finally said it plainly: *"What about writing a screenplay to talk about what mom experienced in the army? I find it odd that it's normal to give a platform like this to an animal before we ever consider giving the same kind of approbation to a black woman."*

My delivery may not have been perfect—but the truth behind it was undeniable. There are countless war stories told from male perspectives. There are stories that honor animals, equipment, and strategy. But there are virtually none that focused on the lived combat experience of a Black woman in modern war, and we are in the 2020s. The reality is this: women had to fight just to be recognized in combat roles. The first woman to officially join the Green Berets was not until 2020. And the first woman to actually complete the Army's Special Forces Qualification Course did so back in 1980—only to be denied graduation because of her sex. Even then, those stories began with white women. Black women and women of color are still waiting to be fully seen.

During that conversation with my husband, something else became clear. He told me—gently and honestly—that he did not know my mother's story well enough to tell it. And in trying to answer his questions, I realized something painful: Neither did I. I knew pieces. I knew artifacts. I knew fragments. But there were gaps only she could fill.

This book exists to close those gaps. It exists to tell the truths she was silenced on; to bring forward the bravery she carried quietly; to name the resilience she embodied without applause; to give voice to thirty-five years of memory, pain, endurance, and survival; and to stand as a mirror for every woman who ever felt invisible in uniform.

My mother refused to disappear. Refused to let them erase her. Refused to let me grow up believing that Black women's stories don't matter. And I refuse to let the world forget she was there. It is not my place to interpret her war. But it is my place to acknowledge the woman who carried it home. The woman who raised me on her own—while carrying memories most of us could never fully comprehend.

What follows is not just a story about the Gulf War or Desert Storm. It is a story about endurance, resilience and survival. Hope versus reality. Identity. And what it means to be a black woman, a soldier, and a mother—all at once. And now, for the first time, she is finally telling it herself. We see you. We honor you. We remember you. You were there. And this story is proof.

Deidra Wilson

Chapter 1: The Recruiter's Promise

I was seventeen years old when I decided to join the Army, and I did not know a damn thing about a lot of things.

What I knew was this: my mother was poor, I was smart enough to dream, and that dreams cost money I did not have. College was not something we talked about in my house, not because my mother did not want it for me, but because wanting something and affording it are two different things. I worked a stay-in-school job doing clerical work for the federal government, making GS-1 pay, which felt like a fortune when I was sixteen, but felt like poverty when I was planning my future.

I knew that staying in DC without a plan meant barely making ends meet—living hand to mouth, paycheck to paycheck, with little opportunity for advancement because I did not have a college degree. Maybe I would go to a trade school eventually, but life would be hard. Really hard.

So, when the Army recruiter showed up at my high school, William McKinley Tech, in the spring of 1989, I listened. Not because I wanted to be a soldier, I did not even know what that meant, really, but because he was selling something I desperately needed: a way out.

৵

The career fair at George Washington University was supposed to be inspiring. That's what my guidance counselor said when she handed me the flyer. "Go see what's out there, Darnnell. Dream big."

So I went. I walked through those pristine halls, passing students who looked like they knew they could afford to go there, and I let myself imagine. I could see it so clearly: me in a white coat, standing in a bright office, a dentist with a good salary and a respectable life. I had done the research. Dentistry paid well, the schooling seemed manageable, and it felt like the kind of career a girl like me could actually complete if given the chance. Then I asked about tuition. The number they gave me might as well have been a million dollars. It was so far beyond anything my mother could afford, so far beyond anything I could imagine scraping together, that it felt like a cruel joke. I stood there in that beautiful building, surrounded by kids whose parents could write those checks without flinching, and I felt something inside me crack. This was never for me. College was never for me. I walked out of that career fair with a pin in my balloon, and I did not tell anyone how much it hurt.

2❧

My mother never talked about college. Not once. It was not that she did not love me or want better for me. She did; but she was a single mother in DC, working hard just to keep my brother Glenn and me fed and housed. College was not in our vocabulary because it was not in our reality. My father was local, but absent; the kind of man who showed up to my high school graduation but was never part of my daily life. I did not know him, and he did not prioritize trying to know me.

So, it was just me, my mother and younger brother; and my mother did the best she could.

I worked because I had to work. I had been working since I was fourteen, earning my own money, and learning how to be independent before I even understood the cost of independence. By the time I graduated high school in June 1989, I had a bank account and a steady paycheck, but absolutely no idea what came next. That's when the recruiter found me.

He was everything you would expect: confident, friendly, incredibly sure that the Army was the answer to every problem I had not even articulated yet. He came to my class one day and stood at the front of the room like he was offering us the world. "You can be all you can be," he said, and I swear to God, I believed him.

He talked about traveling, learning skills, computers, foreign languages and high-tech intelligence work. He made it sound like an adventure; like the military was a place where smart kids like me could thrive. And then, this was the part that hooked me, he said the Army would pay for college. I sat up straighter. "Serve your country," he said. "Learn a trade. Get your education paid for. See the world." It sounded too good to be true, but I was desperate enough not to care whether it was true or not.

❦

I took the Armed Services Vocational Aptitude Battery (ASVAB), and scored high enough to pick any job I wanted. Any occupation, any specialty. The sky was the limit. My boyfriend at the time, who had also taken the ASVAB, did not do as well. He could only choose from a few options: cook, infantry, jobs that did not require such high scores. Not that there's anything wrong with those careers, but for me, I wanted more. I wanted the smart-girl job, the one that felt like it matched who I thought I was. So, I chose Military Intelligence.

I had always loved spy movies, James Bond, covert operations, and top-secret missions. I imagined myself in a sleek suit with a briefcase full of classified documents, attending high-level meetings at the United Nations (UN), and moving through the world with purpose and importance. I genuinely believed I would have that type of life. I was clueless about what I was signing up for.

I believed my family was supportive. No one outright asked, "Are you sure?" or "Do you have other options?" Maybe my mother asked once, gently, but she did not push back. She was probably worried about me leaving home at seventeen, and about me going out into a world where she could not protect me. But I was not worried. I was excited.

Staying in DC felt like stagnation. I could see my future if I stayed: low-level clerical work living paycheck to paycheck, and dreams deferred indefinitely. The military felt like an escape hatch; a way to become someone bigger than the neighborhood I grew up in.

I had hopes. I had plans. I had a vision of myself in six years; older, more educated, worldly, and successful. The kind of woman who had made it out and made something of herself. All I had to do was survive basic training, get through my schooling, and start my adventure. How hard could it be?

ॐ

On August 9, 1989, two months after graduating high school, I raised my right hand at the Military Entrance Processing Station in Baltimore, Maryland, and took this oath:

"I, Darnnell Adams, do solemnly swear that I will support and defend the Constitution of the United States against all enemies, foreign and domestic; that I will bear true faith and allegiance to the same; and that I will obey the orders of the President of the United States and the orders of the officers appointed over me, according to regulations and the Uniform Code of Military Justice. So help me God."

So help me God. I said those words with my whole chest, believing I was crossing into something noble. The Rubicon. The point of no return. I could not have guessed how literal that would be.

I weighed 115 pounds, weak as hell, could not do a single push-up, and had never run more than a block without getting winded. I was a tall, lanky weakling who went to bed at 10 p.m. every night because that is when my body naturally got tired. I was never a night owl. I was not athletic. I was not built for the things they were about to make me do.

I also knew that I was smart, and was determined, and I was sure, absolutely sure, that I could handle whatever came next.

Chapter 1: The Recruiter's Promise

I was wrong about a lot of things. But I was not wrong about one: I was going to survive. I just did not know yet what survival would cost.

RESUME OF
DARNNELL D. ADAMS
706 Jackson St., N.E.
Washington, D.C. 20017
(202) 269-1828

EDUCATION

Backus Junior High 9/83-6/86
South Dakota Ave & Hamilton St. NE
Washington, D.C. 20019

William McKinley SHS 9/86-6/89
2nd and T streets NE
Washington, D.C. 20002

CAREER GOAL

To become a Dentist

WORK EXPERIENCE
5/88 - Present

Student Aide/Clerk Typist
Office for Civil Rights
Department of Health & Human
    Services
330 Independence Avenue S.W.
Washington, D.C. 20201

Duties:
answer and screen incoming calls;
type various documents such as
charts, letters, memoranda, and
reports from rough drafts and
later into final format; run
errands and perform other duties
as assigned

Summer, 1987

Dining Room Attendant/Cashier
Daka Food Services
American History Museum
Washington, D. C.

Duties: clean sections;
operate cash register

SCHOOL ACTIVITIES

Section President
Hy-Flyers Honors Club
Softball Team

Principal Mrs. Betty Tops bestowing my HS diploma June 1989

Darnell can't see

Visiting home during AIT winter break, 1989 (looks like we're at Red Lobster)

Darnell holding Tayvaughn

From left: cousin Allevia, Darnell, uncle Kennny, uncle Victor, aunt Margaret.

Darnell in her favorite position

Darnell fell asleep in the sun

8

Chapter 2: Stripped Down

The day I arrived at Fort Dix, New Jersey, I learned what the military actually thought of me. Nothing.

I was nothing. We were all nothing. A hundred-plus girls fresh off the bus, dragging oversized duffel bags packed like we were going to Disney World. The drill sergeants laughed at us. Actually it was knee-slapping, ha ha hearty ha ha laughing. "Where did you think you were going? Disney World!?" one of them shouted, grinning like we were the funniest thing he'd seen all week. "We weren't even through the gates yet, and they were already stripping us down.

೨೪

The very first thing they made us do, before we ever saw our barracks, before we were assigned bunks, before anything, was march. Fast. With our luggage.

They marched us from the in-processing center to our barracks— about a quarter of a mile, which is only around three blocks, but felt like we were crossing state lines. Later I learned this little ceremony was called the "bag-drag shuffle." My duffel did not have wheels— none of ours did. Mine was brand new, too. I thought I was so sporty and prepared when I bought it just for boot camp. Five minutes into that march, I wanted to throw the whole thing away. It was so damn heavy. Whatever I packed suddenly felt unnecessary for survival. We were carrying raw luggage, old-school style, and there was nothing to do but keep moving.

I was tall and, even with scoliosis, managed to stand straighter than most. But some of the other girls were already at a disadvantage. The shorter ones, the girls with slouched backs, the ones who hunched forward under any kind of weight—they were struggling. I remember one girl in particular, about 4 foot 11 inches tall, wearing oversized glasses and looking absolutely miserable. She was sweating so hard her glasses kept sliding down her nose, and she kept pushing them back up with one hand while trying to wrangle a duffel bag that insisted on slipping off her shoulder. Both her glasses and her bag seemed committed to mutiny, and she was fighting them both at the same time—her short legs working double-time like a Corgi trying to keep pace with the pack.

Bless her heart—those short legs had to practically run to keep up. The drill sergeants had us walking at a breakneck pace, yelling cadence—'LEFT-RIGHT-LEFT-RIGHT!'—so fast it was almost double-time but technically not. Watching her little legs pumping like she was trying to outrun her own luggage… the sight of it cracked me up. I laughed then, and I'm laughing now. I was so grateful in that moment to be tall.

Meanwhile, the drill sergeants reminded us every few minutes that if a single bag touched the ground, we would be dropping for pushups on top of carrying all that weight. So we were speed-walking, hauling heavy duffels, and bracing for punishment the whole way.

By the time I reached my bunk, I was already exhausted and rattled—and we hadn't even started training.

That was the first gut punch.

❦

The second came right after.

We barely had time to drop off our civilian bags before they marched us to in-processing. That's where the real fun began. They issued us everything: uniforms, boots, canteens, sleeping bags, rucksacks, tools, equipment. Everything was green and heavy and smelled like canvas and rubber. We had to carry it all—two full duffel bags, rucksacks on our backs, gear strapped to our bodies—and march it across the post back to our barracks.

If the bag-drag shuffle was a slap in the face, this was a kick in the teeth while we were still down.

By the time I finally reached my bunk for real, I was done. Physically, mentally, emotionally done. And we hadn't even started training yet.

That was the point, I think. They wanted us to understand immediately: You are not special. You are not prepared. You are here to be broken down so we can rebuild you into something useful. And they did. They broke us down.

I was so tired I wanted to cry, but I did not. I could not. Crying would make it worse.

We were also required to learn how to use a buffer machine, a big, heavy, industrial floor polisher that was nearly impossible to control, especially if you were skinny like me. It pulled and jerked and fought you the whole time, and my lower back screamed every time I tried to wrestle it across the barracks floor. But that was not the worst part.

The worst part was the sleep deprivation. We woke up at 4 a.m. Every single day. Sometimes earlier. The drill sergeants loved to burst into the barracks in the middle of the night, banging pots and pans, yelling, flipping on every light, making us stand by our bunks at attention while they "inspected" us for contraband. Contraband was candy and gum and anything that was not Army-issued.

One night, maybe around 2 a.m., the drill sergeants stormed into our barracks screaming at us. We all scrambled out of our bunks in our underwear and t-shirts, standing at parade rest, while they tore through our belongings. And then they found my care package. The one my boyfriend had sent me, full of snacks and candy, that I had stupidly brought up to my room instead of leaving it downstairs in the dayroom like I was supposed to.

The next morning at formation, everyone got smoked; punished with rapid-fire repetition of push-ups, sit-ups and jumping jacks. Everyone that is, except me. I had to stand there and watch while my platoon got smoked so that I could see the cost of my mistake firsthand. I felt awful and it was so embarrassing. It was not a good look to be me that day.

One girl, this ugly chick who looked like Bruce Jenner with stringy, dingy hair and a square manly face, huffed and puffed the whole time, making loud comments under her breath. Loud enough for me to hear. Blaming me. Like I did not already know it was my fault. Like I was not already standing there feeling like shit, watching everyone suffer because of my mistake. She had been trying to act like she was in charge since day one. I already felt terrible. I knew I screwed up. I knew I had gotten everyone punished. But nothing could change it now. What was done was done. And she kept going. Kept making her little comments, her disgusted sighs, her pointed looks. Finally, I snapped. I cursed her out. Right there. I did not like her anyway. She had an ugly face and an ugly attitude, and I was already too exhausted to care about making friends with people who hated me.

Through it all, I learned my lesson. No more contraband. No more mistakes. I did not want to be the reason anyone else suffered.

ஜ☙

The physical training (PT) was relentless. We did PT every morning before dawn, push-ups, sit-ups, jumping jacks and running. My muscles burned. My calves ached. I got shin splints so bad I could barely walk, but we kept marching anyway. Twelve-mile road marches with full rucksacks on soft sand roads that made every step feel like I was sinking into the road.

I was so weak, I could barely do one push-up when I got to basic training, Other recruits had to help me during that first week because I was terrified they'd send me to remedial training for soldiers who could not meet the minimum standards. I refused to fail. Going home was not an option. So, I kept going. I kept pushing. Even when my body wanted to quit, I would not let it.

ஜ☙

The drill sergeants were a special kind of cruel. Drill Sergeant Bodoin had a thick Boston accent and loved to get in your face. Drill Sergeant McDuffy was no better. They yelled at us for everything, standing wrong, looking wrong, breathing wrong. And they did not just yell. They got close in our faces. So close you could see the whole inside of their mouth, smell what they had eaten and feel their spit on your face.

One morning, before dawn, we were standing in formation waiting for orders. It was cold and raining, and we were all just trying to stay still and quiet. I was staring straight ahead, not trying to cause trouble, when another platoon's drill sergeant started pacing in front of us. He was a Black man with dark skin, black-rimmed glasses, and about six feet tall and slender. He reminded me of my granddaddy, who had passed away. My granddaddy was gentle and kind, the type of man who'd give you the second can of Pringles and never raised his voice. I loved him so much. So, I was just looking at this drill sergeant, maybe thinking about my granddaddy, maybe not even really seeing him, when he stopped. He stared right at me. Then he yelled: **"WHAT CHOO LOOKING AT, BUG EYES?!"**He really punched the word *bug*. Drew it out. He made sure everyone heard.

"Every time I look up, you looking at me!" he said. I did not say a word. I did not move. I just stood there, humiliated, while he stared me down and the rest of the platoon stayed silent. That's what drill sergeants did; they found your weakness, your face, your body, your fear, and they used it to break you. I was not Bug Eyes before that moment. Thankfully, I don't think anyone knew he was talking to me. We could not turn our heads while in formation. Whether they knew, I can't say. But thankfully, nobody ever mentioned it.

ॐ

The hardest thing I had to do in basic training was qualify on the rifle range. I failed the first time. I had 20/20 vision so I believed that would make it easy. But lying in the prone position on wet, muddy ground, trying to hold my breath steady, looking through that narrow sight alignment was so much harder than I expected.

It was a gray, cloudy day. It might have rained. I was one of the last two people still trying to qualify. Everyone else had qualified and left. Even the short, quiet girl with the Coke-bottle glasses qualified first and got the best score. I was so discouraged. But I kept trying. I stayed out there all day, shooting, adjusting, breathing, praying I would hit the target. And finally, I did. I earned my **Sharpshooter** badge. If I hadn't passed, they could have held me back, delayed my graduation and sent me to remedial training, but I made it.

Grenade training, though? That was easy. All I had to do was simulate pulling the pin and throw the grenade as far as I could toward the target. That's it. No measuring distance and no aiming. Although, I threw like a girl, I passed on the first try and earned my **Expert** badge. I was proud of that.

ॐ

The gas chamber was a different kind of hell.

They put us in a small, rectangular building with no windows and one door. We wore our uniforms and had our nuclear, biological, chemical, NBC masks strapped to our thighs. The drill sergeants explained that we were about to be gassed with tear gas, and our job was to stay calm and put on our masks.

This was training for a real-life scenario, what to do if we were exposed to poison gas in a combat zone. When then drill sergeants left, the gas started. It had a chemical smell that burned the back of my throat. My eyes started watering and burning. I panicked just like everyone else. My hands fumbled with the mask, trying to get it over my head, trying to seal it to my face by taking a deep breath.

But taking that breath hurt. The gas that was already in my mouth went down into my lungs, and my body wanted to reject it, coughing, gagging, running.

Private Burris absolutely lost it. She started flailing, screaming, throwing herself backward. **"LET ME OUTTA HERE! I CAN'T BREATHE! OPEN THE DOOR, DRILL SERGEANT, OPEN THE DOOR!"** She was crashing out. Full-on panic attack. The other soldiers tried to help her, but they were no better off. Those of us who got our masks on were still uncomfortable, still scared, but we were out of danger. Finally, the drill sergeants opened the door and pulled Burris out. She collapsed on the ground, sobbing, coughing, gasping for air. We all thought it was hilarious. Burris became legendary after that, the girl who went crazy in the gas chamber.

But honestly, I understood. The gas chamber was terrifying. I was just glad I kept my head.

ॐ

I did not make a lot of close friends in basic training. Not at first. I was cool with the other Black girls in my platoon, but I was not trying to build lasting relationships. I just wanted to survive and get out.

There were some memorable people in my class: **"Lil Bit"** (Private (Pvt.) Browning) was maybe five feet tall with a face like a cute little teddy bear. She had a soft, high-pitched, country voice. She was from Louisville, Kentucky, and always seemed to have a frown on her face, probably because her feet were covered in blisters from those damn boots.

Pvt. Sanora Bond was short and quiet, until one of the female drill sergeants got in her face and started yelling. Bond yelled back. The male drill sergeants laughed and pointed out that Bond and the drill sergeant looked like identical twins. After that, Bond and the drill sergeant became friends.

Pvt. Tanoah Bond was memorable only because her name was almost identical to Sanora's. The Bond girls.

Pvt. Burris, the gas chamber girl, was tall, slim, curvy, with a raspy voice and a thick country accent. She was older than most of us and constantly talked about douching. Every day. She said she did not care what the doctors said, she did not feel clean unless she douched.

Pvt. Flowers reminded me of my mom. She was tough, from Shreveport, Louisiana, and nobody messed with her because she would speak up for herself with the quickness.

Pvt. Anglin looked like a female Eazy-E. She had a Jheri curl and a pimp walk and thought she was going to run the platoon with her masculine energy. She quickly found out that nobody cared. Most of the other girls were inner-city tough girls who weren't afraid to throw hands.

Pvt. English seemed sweet and innocent, maybe a little goofy. One evening while we were waiting to shower, she asked if she could wash my back and I could wash hers. I was confused. I said no. I could not figure out if she was hitting on me or, just being friendly. Looking back now, I think she might have been gay. But I was too naïve to understand it then.

ॐ

There were funny moments, too. Embarrassing moments and there are some things you just don't forget. Some moments are so absurd, so uncomfortable, so deeply *what the hell is happening right now* that they burn into your memory forever. Like the night my bunk buddy washed her coochie in our tent—that's one of those moments. We were on a field exercise, bivouac they called it, which meant we were camping out in the wilderness in these tiny two-person pup tents. After a brutal 12-mile road march, I was exhausted. It was dark, cold, and all I wanted to do was crawl into my sleeping bag and pass out. Every pair of bunk buddies was assigned a guard duty rotation. One person would sleep while the other stood watch on the perimeter, then we would switch. My bunk buddy was supposed to go first, so I figured I would get a few hours of rest before my shift. But she had other plans.

She decided, right then, right there, in our cramped little tent with me lying inches away, that she needed to wash herself. Her *whole* self. Specifically, her coochie. She took off her pants, squatted down, and used her helmet as a makeshift sink. Just started going to town, washing her naked private parts while I lay there frozen, trying not to breathe, trying not to *see*, trying to figure out how to escape this situation without crawling over her bare ass.

I could not take it. I held my breath, scrambled out of the tent as fast as I could, and went to find the other soldiers so I could tell them what the hell had just happened.

We all lost it. We were laughing so hard we could barely stand. How bold do you have to be to strip down and wash your coochie in someone's face? In a small, shared tent; In the middle of the night;

with no warning to your bunk buddy? It was crazy. Insane. The kind of thing you cannot make up.

But that's basic training. That's the military. You're thrown together with all kinds of people; people from different backgrounds, different levels of hygiene awareness, different ideas about personal space, and you just have to deal with it. So, I dealt with it. I made sure I was nowhere near that tent when she finished her…ablutions. And I made damn sure I never agreed to share a tent with her again.

≈

By the time I graduated basic training in October 1989, I had been broken down and rebuilt the same way I learned to dismantle and rebuild my M16 A1 rifle: with expertise. I learned how to follow orders, how to push my body past its limits and how to survive on little sleep and less kindness. I also learned that the military did not care about me. Not really. I was a body, a number, and a trainee who either made it, or did not make it. But I made it. And I told myself: the worst is over. Advance Individual Training (AIT) will be easier. I was wrong about that too. By the end of basic training, I had learned a lot of lessons. Some practical, some painful, some just plain weird.

I learned how to do push-ups, how to march with a heavy rucksack, how to shoot a rifle and throw a grenade, how to survive on Meals Ready to Eat (MREs) and little sleep and even less privacy.

I learned that the drill sergeants did not hate me personally, they hated everyone. That was their job. I learned that I was stronger than I imagined. That I could push my body past what felt like its breaking point and keep going. And I learned that no matter how bad things got, there was always something absurd enough to laugh about later.

All of it—the candy that got everyone punished, Private Burris losing her mind in the gas chamber, my bunk buddy washing her coochie in my face, the hours I spent on that range until I finally qualified with the M16—these were the moments that shaped who I was becoming.

After graduation, I became a holdover. My AIT class did not start until after Christmas, so I had to stay at Fort Dix for about a month while everyone else left for their next duty stations. It was strange. I had more freedom, and more downtime. I was not getting screamed at every five minutes. I could actually breathe. That's when I met Alberta McGee.

Alberta was soft-spoken, warm and sweet as apple pie. She had a heavy southern drawl and the kind of gentle presence that made you feel safe just being around her. We would go to chow together, walk around the post, talk about where we were headed next and what we hoped our lives would look like.

I liked Alberta a lot. She was one of the first people in the military who felt like a real friend, not just someone I was surviving alongside. We lost touch after we both left for AIT. I never saw her again. But I still think about her sometimes, and I hope she's doing well.

꒰๑꒱

Basic training did not make me love the military. It did not inspire me or fill me with pride. But it did teach me something important:

I am a survivor. I learned that could endure humiliation, exhaustion, fear, and absurdity, and come out the other side still standing. I earned my sharpshooter badge. My expert grenade badge. I made it through the gas chamber without losing my mind. I graduated. And now I was headed to AIT, Advanced Individual Training, where I learned my actual job: Military Intelligence. Electronic Warfare, Signal Intelligence. I was going to Fort Devens, Massachusetts, in the dead of winter, to learn skills I barely understood. But I was hopeful and I was determined.

I convinced myself that the worst was behind me. I was completely unaware what was awaiting me in cold New England.

Monday Oct. 3

Dear Ma,

I just found out that our pass on the 11th doesn't start until 1700 hrs. 5:00 pm. And you have to check me out. So please be here before five. I'm sure you will be able to visit me but I just won't be able to leave.

My Barracks are located on Augustus Street.

I'll try to draw a picture and explain better because theres no intersecting street to let you know where on Augustus I stay. If you can find the PX and the MCC Phone Center you'll find me.

— over —

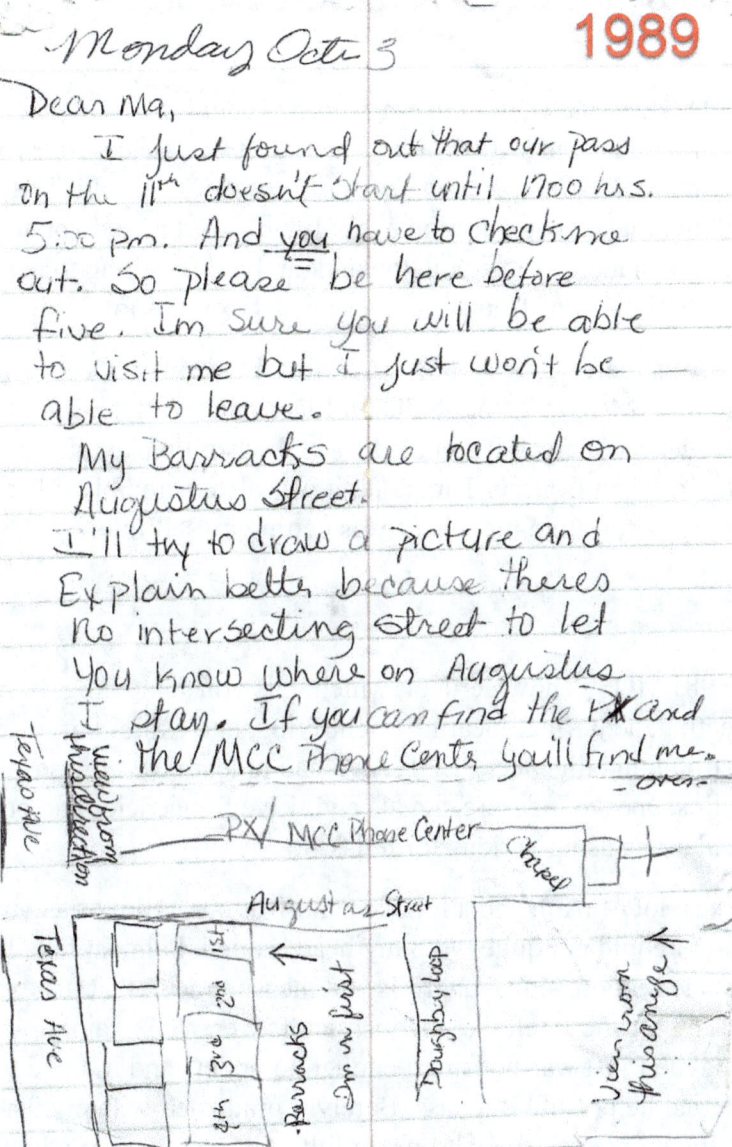

Darnnell's letter to her mother about basic training graduation

Chapter 3: AIT - Learning the Craft

Fort Devens, Massachusetts, in January is cold in a way that DC girls may not understand. It was not just cold. It was *freeze-your-face-off, can't-feel-your-fingers, why-did-I-join-the-military* cold. The kind of cold that makes you question every life choice that led you to this moment, standing outside in the dark at 4 a.m., waiting to march to a chow hall that was fifteen minutes away through snow and ice.

But I was there. January 1990. AIT for my, Military Occupational Specialty (MOS) 98 J or 98-Juliet. Electronic Warfare Signal Intelligence Analyst. It sounded so cool when I picked it. So high-tech. So Jason Bourne. Turns out, it was also incredibly difficult, and it never occurred to me what I was getting myself into.

⁊❧

The 98J MOS is now defunct, which tells you something. It involved detecting, acquiring, locating, identifying, and exploiting foreign electronic intelligence, ELINT. We had to learn how to use oscilloscopes to measure signals and wave frequencies and perform calculations using various formulas.

It was a lot of math and a lot of memorizations. A lot of technical jargon I could not quite wrap my head around. I almost failed. There was one section, signal analysis and measurement, that just was not clicking for me. I felt like the instructors weren't explaining it well, or maybe I just was not smart enough to understand it. I had to get remedial help. But I did pass. Barely. I graduated with my 98J Certification, even though I never fully felt confident in what I was doing.

I told myself I would figure it out once I got to my unit. That real-world experience would make it all make sense.

Spoiler: it did not.

⁊❧

Fort Devens was not all bad, though. There were moments of lightness, even joy, tucked between the brutal cold and the technical confusion. Like the morning my friends and I slipped on ice. It was one of those o'dark-thirty mornings, pitch black, freezing, probably below zero. My three cohorts, Dana Hale, Taria Wright, and Satonya Petty and I were walking from our barracks to the chow hall, bundled up in our winter Battle Dress Uniforms (BDUs), scarves, hats, and gloves. We took a shortcut through the woods, a path we had taken a hundred times before. But that morning, the entire path was covered in black ice.

As soon as we took our first step onto the trail, all four of us started sliding. We weren't just slipping, we were *careening* out of control, arms flailing, trying to grab onto each other as we slid down the entire hill in one tangled, screaming heap.

"Woahhhh!!" we yelled in perfect harmony, like a terrible a cappella group. We landed at the bottom of the hill in a pile, like felled bowling pins. And for a second we just sat there on our bums, stunned. Then we fell out laughing. Laughing so hard we could not breathe. Our arms had automatically interlocked as we slid, like some kind of involuntary group hug. The whole thing was so ridiculous, terrifying and funny that we could not stop laughing. We laughed all the way to the chow hall. This was one of my favorite memories from AIT. Not the classes. Not the cold. Just that stupid, joyful moment of sliding out of control with my friends down an ice-slicked hill, without getting even a scratch then laughing uncontrollably when we realized we were all unharmed and survived that scary ride down.

⁊❧

There was also the blanket party. A blanket party is a military term for a form of unofficial discipline. It's when a group of soldiers conspire to physically punish another soldier for not conforming to expected standards, usually things leadership didn't see because it happened in the barracks, away from official eyes. Traditionally, they would throw a blanket over the person's head so they could not be identified by their attackers, then beat them. This party was not quite that violent, but it was a blanket party, nonetheless.

There was a girl in our barracks who refused to shower. She would come back from PT, from running, from field exercises, and she just wouldn't bathe. Ever. The smell was unbearable. We slept in an open-bay setting where all the bunks were visible, so her stench permeated the entire floor. We tried talking to her. Gently at first. "Hey, you know, it's important to shower after PT. We all do it." She ignored us. So one night, the other soldiers on her floor decided enough was enough. They grabbed her, carried her to the showers, and physically washed her while she fought and screamed.

I did not participate in the blanket party. My bunk was on the upper floor, and I was not in the barracks most weekends anyway—I was out with my boyfriend. But I was privy to the planning, and in my opinion, I did not blame them. Sometimes when someone doesn't listen, doesn't follow basic hygiene that affects everyone, soldiers take matters into their own hands.

ॐ

My boyfriend was another trainee in my AIT class, and we started dating a few months into 1990. He had this old beat-up car he called Batmobile, and every Friday after class, we would leave post and drive to Leominster, Massachusetts, where we would stay at a motel for the weekend. It felt like we were living in our own little apartment. Just the two of us, away from the barracks, away from the cold, away from the drill sergeants and the constant pressure. We would order takeout, watch TV, sleep in, pretend we were normal people with normal lives. It was one of the best part of AIT. Until I lost his car keys.

ॐ

I had a hole in the pocket of my long black wool winter coat, and I did not realize it. One snowy day, I had his keys in my coat pocket, and they must have slipped through the hole and fallen into the snow. We looked everywhere, around the barracks, in the parking lot, all over the areas we had been walking. Nothing.

My boyfriend was calm about it, but I felt terrible. We had to start taking taxis to our motel on the weekends, which was fine, but it was not the same. And getting food afterward was tricky. We would either have to stock up on snacks beforehand or order delivery. Months later, after the snow melted, we found the keys right where we had been looking all along. We were so happy. The Batmobile was back in business.

Darnnell at Fort Devens

ॐ

The funniest thing that happened during AIT involved one of our instructors and the MSQ1O3 Charlie, the massive signal acquisition vehicle, we used for training.

This instructor was a Black guy with a lisp who sounded like Daffy Duck when he talked. He was goofy and quirky, but always nice. One day, he was demonstrating how to drive the MSQ1O3 Charlie, a very expensive piece of equipment, and he impressed things such as the importance of spatial awareness, knowing the vehicle's height, and being careful around low structures. Well guess what, *he* forgot to lower the antenna mast and he drove the vehicle straight into the hangar opening and smashed that piece of expensive equipment on the roof.

We all just stood there, mouths agape, and then we burst out laughing. We could not help it. He was stuttering and stammering, looking embarrassed and nervous, but it was the highlight of our entire training. After that, I felt a little better about not understanding signal analysis. If the *instructor* could crash a vehicle, he was supposed to be an expert on, maybe I was not as hopeless as I thought. But AIT was not all fun and games. There was Drill Sergeant Munster.

Munster was in my chain of command, and he was foul. The worst kind of man, a predator with power. He was openly having a sexual relationship with a young female soldier in his command named Molina Gonzalez, and everyone knew about it. He did not even try to hide it. He also made sexual advances towards me. Constantly.

He never touched me, but I made sure I was never alone with him. His overtures were relentless. The way he looked at me. The things he said. The way he would corner me physically, testing boundaries, seeing how far he could push. I told my boyfriend about it, and he hated him. We all did.

After I graduated, Munster got kicked out of the Army. His relationship with Gonzalez finally became known to his superiors, and they discharged him but at that point, the damage was already done. Munster's behavior was the first time I truly understood that men in positions of authority over me could be dangerous and that they could use their power to violate, manipulate, and exploit. It was a lesson I would carry with me for the rest of my life. And it was a lesson I would learn again, much more painfully, in the desert.

ᏄᎿ

Despite everything that I had experienced… the cold, the confusion, the harassment, I graduated from AIT in April 1990.

I was now officially a 98 Juliet; an Electronic Warfare Signal Intelligence Analyst, and I had my orders, and that part was a whole other disaster.

NAME: ADAMS DARNNELL
SSN: 577

MILITARY COURSE COMPLETIONS

COURSE NUMBER: 750-BT
TITLE: BASIC COMBAT TRAINING
      BASIC TRAINING

ACE GUIDE NUMBER
AR-2201-0399

LOCATION: US ARMY TRAINING CENTER FT DIX, NJ
DATES TAKEN: 08/18/1989-10/12/1989
DESCRIPTION:
UPON COMPLETION OF THE COURSE, THE RECRUIT WILL BE ABLE TO DEMONSTRATE GENERAL KNOWLEDGE OF MILITARY ORGANIZATION AND CULTURE, MASTERY OF INDIVIDUAL AND GROUP COMBAT SKILLS INCLUDING MARKSMANSHIP AND FIRST AID, ACHIEVEMENT OF MINIMAL PHYSICAL CONDITIONING STANDARDS, AND APPLICATION OF BASIC SAFETY AND LIVING SKILLS IN AN OUTDOOR ENVIRONMENT. INSTRUCTION INCLUDES LECTURES, DEMONSTRATIONS, AND PERFORMANCE EXERCISES IN BASIC MILITARY CULTURE/SUBJECTS, INCLUDING MARKSMANSHIP, PHYSICAL CONDITIONING, FIRST AID, AND OUTDOOR ADAPTATION/LIVING SKILLS.

ACE CREDIT RECOMMENDATION:
IN THE LOWER-DIVISION BACCALAUREATE/ASSOCIATE DEGREE CATEGORY, 1 SEMESTER HOUR IN PERSONAL PHYSICAL CONDITIONING, 1 IN OUTDOOR SKILLS PRACTICUM, 1 IN MARKSMANSHIP, AND 1 IN FIRST AID (10/00).

COURSE NUMBER: 233-98J10
TITLE: NONCOMMUNICATIONS INTERCEPTOR/ANALYST

ACE GUIDE NUMBER
AR-1715-0710

ELECTRONIC WARFARE (EW) SIGNAL INTELLIGENCE (SIGINT) NONCOMMUNICATIONS INTERCEPTOR
LOCATION: INTELLIGENCE SCHOOL FT DEVENS, MA
DATES TAKEN: 11/27/1989-04/03/1990
DESCRIPTION:
UPON COMPLETION OF THE COURSE, THE STUDENT WILL BE ABLE TO USE TECHNICAL REFERENCES AND EQUIPMENT TO SEARCH FOR SELECTED CLASSES OF INTERCEPTED NONCOMMUNICATION ELECTRO-OPTIC SIGNALS AND RECORD, ANALYZE, AND IDENTIFY THESE SIGNALS. LECTURES, DEMONSTRATIONS, AND PERFORMANCE EXERCISES COVER MAP READING; DATA AND SIGNAL ANALYSIS; BASIC ELECTRICAL AND COMMUNICATIONS THEORY; REPORT WRITING; OPERATION AND MAINTENANCE OF RADIO COMMUNICATIONS EQUIPMENT; SOVIET AND NORTH KOREAN DEFENSE SYSTEMS; AND THE IDENTIFICATION, COLLECTION, AND REPORTING OF INTERCEPTED NONCOMMUNICATIONS SIGNALS.

ACE CREDIT RECOMMENDATION:
IN THE LOWER-DIVISION BACCALAUREATE/ASSOCIATE DEGREE CATEGORY, 2 SEMESTER HOURS IN INTRODUCTION TO RADAR SYSTEMS AND 2 IN PERSONNEL MANAGEMENT (6/91).

MILITARY OCCUPATIONAL SPECIALTIES HELD: 98J10 PRIMARY 02/1990-07/1992
      98J10 DUTY
SQT (THRU OCT 1991)/SDT (NOV 1991 THRU FEB 1995) TAKEN: NONE

MILITARY OCCUPATIONAL SPECIALTY GROUP: 98J (PRIMARY)
TITLE: NONCOMMUNICATIONS INTERCEPTOR/ANALYST

ACE GUIDE NUMBER:
MOS 98J-003

DESCRIPTION OF 98J10:
ERECTS NONCOMMUNICATIONS COLLECTION SYSTEMS; CONDUCTS SEARCH FOR SELECTED CATEGORIES OR CLASSES OF NONCOMMUNICATIONS OR ELECTRO-OPTICAL SIGNALS; USES TECHNICAL REFERENCES AND EQUIPMENT TO RECORD AND PERFORM PRELIMINARY ANALYSIS AND IDENTIFICATION OF INTERCEPTED SIGNALS OF INTEREST; DETERMINES LINE BEARINGS OR INTERCEPTED SIGNALS WITHIN SPECIFIED LIMITS; REPORTS ACQUIRED INFORMATION; POSTS ENTRIES AND KEEPS LOGS; OPERATES COMMUNICATIONS EQUIPMENT FOR INTELLIGENCE REPORTING AND COORDINATION.

Chapter 4: Cold Welcome

During Jay school, the best part of our day was not the code tables or the drills — it was when our instructors told stories about real duty stations.

One Black instructor had the class hanging on every word when he described his time at a place called "Wild Chicken."

"Why do they call it that?" we asked. "Where is it?"
He would shake his head in dejection and say, "That's the problem. It's in the middle of nowhere. Freezing cold, nothing to do, and too far from civilization to complain."

We all laughed and promised each other we would never end up there.

So when my orders arrived showing Wildflecken, Germany — my stomach dropped. Wild Chicken. The very place I had sworn off. It was supposed to be somewhere near the Fulda Gap, up in the mountains, remote and bitterly cold. The name alone sounded like punishment. And for a young Black woman in Military Intelligence — the only one in my 98J class — it did not feel like the European adventure I was hoping for. It felt like exile. I was crushed.

꩜

When I landed in Frankfurt that April morning in 1990, I was told to take the bus to Wildflecken the next day at 0700. That night, I stayed in a transient barracks near the processing station, staring at the ceiling and crying myself hoarse. The echo bounced off the empty bunks like confirmation: I was truly alone.

The next morning, I was supposed to get on the bus north.
but I didn't. I boarded one going in the opposite direction. One of my instructors had once joked that if you did not like your assignment, you could just "get lost" on the right bus and end up somewhere

better. It was not supposed to be literal advice, but in that moment, I had nothing to lose.

Before I left, I called home. My aunt Patty answered at my grandmother's house. I tried to be strong, but when she asked how I was, I broke. "I thought Germany was supposed to be nice!" I sobbed, sounding like Lucy Ricardo on a bad day. She tried to calm me, but there was not much she could say. So, I wiped my face, grabbed my duffel, and got on the bus heading south — toward Lüdwigsburg.

᠄

When I arrived at the 511th Military Intelligence Battalion, twenty minutes from Stuttgart, I played it cool. Walked in like I belonged there, like my orders had always said Lüdwigsburg. The soldiers were friendly enough, and I started to relax. Then someone mentioned that Stuttgart's barracks were even nicer — co-located with the Air Force. Apartment-style, clean, modern.
And that's when my greed kicked in. I figured if I had come this far, why not go all the way? So later that day, I caught another bus and headed straight for Stuttgart. The noncommissioned officer in charge (NCOIC) there gave me temporary quarters in the Air Force-style barracks — private rooms, balconies, carpeted floors. For about five minutes, I allowed my immature mind to think I had pulled off the greatest hustle in Army history. Then the phone rang. The First Sergeant from the 511th had found out I had gone missing. She drove straight to Stuttgart, found me, and hauled me back to Lüdwigsburg herself. And she was furious.

In her office, she locked me in parade rest —both arms behind your lower back, feet shoulder width apart and eyes locked straight ahead, and she let me have it. Her words hit like rifle fire — sharp, precise, unrelenting. I stood there, staring straight ahead, swallowing every bit of it. From that day forward, she saw me as trouble. A slick, fast-talking private who thought she was too good for orders.

And maybe I was too good for those orders or maybe I was just immature. Because underneath the attitude was a scared kid who did not know how to ask for help. I was not trying to be rebellious — I was desperate not to disappear into some freezing mountain post of which no one had ever heard.

So yes, I disobeyed orders. And yes, I paid for it. but I also landed in Lüdwigsburg — not Wildflecken, not Stuttgart, but somewhere in between. And as it turned out, Coffey Barracks was not bad at all. There was a bowling alley next door that sold cold beer and the best sourdough pizza I had ever tasted. A snack bar that served taco salads piled high with cheese, tomatoes, and sour cream in a crispy edible bowl. I joined a bowling league. Learned proper form. Discovered white soldiers could line dance as hard as any crowd back home.

It was not the glamorous assignment I had imagined.
But it was the first place that started to feel like mine.

Chapter 5: Lüdwigsburg Life

Germany and Coffey Barracks as it turned out, was pretty great. Not at first. Not while First Sergeant was angry with me. Not while I was trying to prove that I was not the screw-up she thought I was when she first met me.

But eventually, I settled in. And once I did, I realized that Lüdwigsburg was everything I had hoped for in a base assignment.

The WWII barracks weren't fancy, but they were livable. I had a two-person room with twin beds on opposite sides, tall crank windows that opened wide, and a roommate who was rarely there because she had a fiancé off-base. I had a small TV and a boombox cassette tape player. I loved listening to my favorite artists: Mariah Carey, Anita Baker, or watching movies in my friends' rooms. The white guys in my unit had huge VHS movie collections, and since they watched movies I had never seen, everything felt new and exciting. I cooked in my room on a two-burner hotplate. Pan-seared pork chops. Simple meals that tasted like home. Other soldiers would tell me they could smell my food cooking, and it smelled so good.

We did not have cell phones in 1990. If you wanted to call home, you had to walk down to the CQ (Command Quarters) office where there was one phone on the floor that soldiers could use on weekends. Sometimes people would call that phone, and if you happened to be in the hallway, you could answer it and go tell the person they had a call. Or you could just ignore it and keep walking. It was a simpler time.

When I first arrived in country, the unit arranged for me to take introductory German lessons at Robinson Barracks. A German national taught the class—someone who actually spoke the language, not just an American soldier who'd picked up a few phrases. She taught us key phrases we would need to survive off-post. Was ist los? (What's going on?) Wo ist die Bahnhof? (Where is the station?)Wie viel? (How much?) And taught us how to count to ten: eins, zwei, drei, vier, fünf, sechs, sieben, acht, neun, zehn. I practiced those phrases constantly. Used them every chance I got. I wanted to feel like I belonged there, like I could navigate this foreign country with some competence. And happily, It worked. Learning even that little bit of German made me feel less like a tourist and more like someone who was actually living there.

The food in Germany was incredible. On post, my favorite meals were the bowling alley sourdough pizza and the taco salads from the snack bar. Sometimes I would eat at the chow hall, and I loved getting Shirley Temples at the post nightclub, but off post– That's where the magic happened.

Gyros! Germans made the best gyros I have ever had in my life. They slow-roasted huge lamb shanks on a spit until the meat was tender and juicy, then carved it onto pita bread and stuffed it so full it barely folded. They'd top it with tzatziki sauce, but I never wanted it. The meat was so flavorful on its own, I did not need anything else. I could have eaten those gyros every single day.

Ooh, and fried duck from this authentic Chinese restaurant. I have never had duck that good since. The tender cuts of meat were crispy on the outside and moist and flavorful throughout, perfectly oven fried, with no excess fat. I would leave there so stuffed I had to unbutton my pants just to breathe.

Henchen legs, deliciously crispy baked chicken legs with seasoning that was finger-licking good. The skin was juicy and crispy at the same time. I don't know what kind of crack they put in that chicken, but it was incredible.

Brötchen, hot, buttery bread sold by a little old German lady who drove around in a slow-moving truck with a single ding of a bell. We called her the brötchen lady, and she was like the ice cream man for soldiers. Everybody would come running when they heard that bell.

Darnnell at the snackbar at Coffee Barracks

Gasthaus meals, grilled beef or pork from these inconspicuous little inns you'd never notice from the outside because it looked like a little hole-in-the-wall cobbler's home. But inside, they had the most banging food. You could always get a perfectly grilled pork chop or steak with golden, crispy pommes frites (French fries) and a "Coke mit Eis" (with ice)—you had to specify, or you'd get no ice, which Germans did not normally prefer with cola.

And German ice cream was so delicate, rich, refreshing, with crisp chocolate wafers used to scoop the ice cream in the cup. I especially loved their chocolate ice cream. It tasted homemade in the best way.

I learned how to navigate the German bus and subway system, the Bahnhof. I would walk to the nearest bus stop in the neighborhood nearest Coffey Barracks, take the bus downtown to the town center, and spend the day shopping or eating lunch at a Gasthaus by myself. I loved those solo outings. I felt independent. Grown-up. Like I was living the European adventure I had dreamed about. I also traveled.

I went to Bodensee on a weekend trip with some barracks mates who had cars. They would invite me along, and I jumped at the chance. Bodensee was this lovely, picturesque town on Lake Constance, and it felt like something out of a fairy tale. I went to the Black Forest and bought a cuckoo clock. It seemed like the thing to do—you go to the Black Forest; you get a cuckoo clock. I still have it thirty-five years later, though I've only used it maybe once. It's annoying as hell honestly. But I kept it because it reminds me of that day, of being young and free and buying souvenirs for a future I imagined.

I explored on my own too and I would take local buses into town, hop on the Bahnhof to other cities, or grab a taxi if I was feeling fancy. I was confident, curious and brave. I wanted to see everything. But the place I loved most was Robinson Barracks.

RB, as we called it, was in Stuttgart—pretty far from Lüdwigsburg—but it was worth the trip. It was the mecca for the military in our area. The PX there was huge, stocked with everything you could want, and it reminded me of Landover Mall back in Maryland.

Landover Mall was one of my favorite places growing up. It is closed now, but back then it was this bustling, vibrant place where I would go with friends or family, just walking around, window shopping, dreaming about the things I would buy when I had money.

Robinson Barracks felt like that. Familiar. American. A little piece of home in the middle of Germany.

I would take the military shuttle bus to get there. The shuttle circulated through all the kasernes in the Stuttgart area—Coffey, Flak, Patch, Kelley, Panzer, Robinson. It took forever to make all the stops, but I did not mind. I would make a whole day out of it. And when I finally got to RB, I would wander the PX for hours.

That's where I bought my first boombox with a cassette and CD player. CDs had just come out, and I was so proud of myself. I had my own money. I was living in the barracks, so I did not have to pay rent, which meant I actually had pocket money for the first time in my life. I could buy things I wanted, not just things I needed. That boombox felt like freedom.

I usually went to RB by myself. So I shop or browse at my own pace, eat where I wanted, stay as long as I wanted. I loved being in the sanctuary of my own solitude. But I had friends too. Real ones. Dawn J., from my unit in the 511th, Meluxana, Christine and Mike Heflin, who I had known since Fort Devens—they'd met during AIT, fell in love, got married, and they are still together to this day.

Dawn was a no-nonsense white girl with a thick Boston accent and the kind of presence that made you feel instantly safe. She was authentic, funny, and the coolest girl ever. I gravitated toward her because nothing about her was fake. I believe she and Meluxana were friends before I met her. We were all stationed at Coffey Barracks together, though Dawn was in my unit and Meluxana was assigned to another.

The three of us had the best times—hanging out in each other's rooms, cracking jokes, and going out on "the economy," which was just our way of saying we were leaving post to explore Germany. On weekends we would grab a bite to eat or just wander around the base. The bowling alley, the snack bar… I was always up for a good meal, so if food was involved, I was there.

And Meluxana—was hilarious. He was from Thailand and swore up and down that Thailand had the prettiest women in the world. I believed him. He was so proud of his heritage, and he carried himself with this joyful confidence. He was another person who was safe, fun, and just easy to be around.

Meluxana dancing in my room

And then there was Sonny Quest. One evening, a knock came at my barracks door. A female soldier I did not know stood there. "There's a guy downstairs who wants to meet you," she said. "Who?" I asked. She told me his name—I think it was his last name, maybe his first. I don't remember now which it was. "I'm in a relationship," I told her. "My boyfriend's stationed in Turkey. "She waved that off. "Just go talk to him. He's waiting on the steps outside." I don't know why I went. Maybe curiosity. Maybe because she was so insistent. But I walked downstairs. He was waiting on the outside steps of the barracks, leaning against the railing like he had all the time in the world. White guy, blond hair, maybe five-ten or six feet, with a stout, solid build—about 160 pounds. He had a baby face, not rugged, more like a farm boy who hadn't quite grown into himself yet. But his eyes were vividly blue—the kind you noticed. We talked briefly. He was confident—maybe too confident. At some point in the conversation, he told me he was "God's Gift to Black Women. "GGBW, he called himself. I remember thinking he was pushy. Too eager. Like he expected me to be impressed. "I'm not interested," I told him. He took it in stride, but I could tell he was not used to hearing no.

He was an 11bravo (B)—an infantryman in Foxtrot Company. His building was at the rear of Coffey Barracks, just like mine. Different company, different mission, but the same small post where you'd see the same faces over and over. After that, I would see him occasionally at the post nightclub when my friends dragged me out for country line dancing nights. I would sit at the bar drinking Shirley Temples with extra cherries, watching everyone do those fancy steps I could never get right. I tried—God knows I tried—but I was awful at it.

Quest would sit at the bar with me sometimes. We would chit-chat. Nothing serious. Just two soldiers killing time while everyone else danced. I did not think much of it then. Just another guy. Another conversation. Another night at the club. I had no idea that months later, when we deployed to Desert Storm, this same infantryman would somehow end up on my team under Strychland. An 11B attached to a Military Intelligence unit. I don't know how he worked that out, but looking back now, I wonder if it was orchestrated all along.

꩜

But all in all, I was happy. Really, genuinely happy. For a little while anyway. In Germany, I had gyros and brötchen and trips to Bodensee. I had my boombox and my little room and my solo adventures on the Bahnhof. I had moments where I felt free. And I held onto those moments as long as I could.

Chapter 6: Riding the Beast

The 511th Military Intelligence Battalion was a tactical unit, not a strategic one. That meant we weren't doing office work in top-secret facilities like the NSA. We were doing field training exercises, lots of them. Out in the cold, muddy wilderness, setting up equipment, running drills, and living in tents.

This was not what I had envisioned when I chose Military Intelligence. I was expecting to wear my Class-B uniform with either pressed slacks or a skirt and my army issued regulation pumps or low quarter loafers while I was attending those high-level meetings, INDOORS. Living my ultimate smart-girl dream. Instead, I was roughing it outdoors. But there was one part of field training I absolutely loved: the MSQ103 Charlie maneuvers.

Oh! My! Goodness! The MSQ103 Charlie is a beast! It is a vehicle with tracks like a tank that could tear up the ground and crush small trees under its treads. It was our electronic warfare signal intelligence tracking vehicle, and it was the coolest thing I had ever seen or driven.

The first time we took it out into the field, I was giddy with excitement. We rolled out in a unit convoy, armored personnel carriers, deuce-and-a-halfs, Humvees, CuCV (Cuck Vee - Commercial Utility Cargo Vehicle) and our MSQ103 Charlie. My team leader, Sergeant Shane, and my teammate, Specialist Burns, and I all climbed into the vehicle. Burns drove, and we took off into the woods. It was incredible.

The track vehicle devoured everything in its path. Skinny trees– crushed. Tall weeds and vines– flattened. We were going up and down like a roller coaster on the uneven terrain, and I was laughing like a kid at an amusement park, but it was dangerous as hell. We weren't wearing seatbelts, I don't think there were seatbelts. The engine was loud, and we were bouncing around inside this metal beast, but I did not care. It was the most fun I had ever had in the Army. For a brief, shining moment, I felt like I had picked the right job. Like maybe this whole Military Intelligence thing was going to work out after all. And then we went to Hohenfels.

Hohenfels Training Area was in the mountains—high elevation, the kind of place where the air felt thinner and the views stretched forever. It was one of the Army's main training areas in Germany, and we were there to do serious work. The MSQ1O3 Charlie came with us, of course. That beast of a vehicle went everywhere we went.

But at Hohenfels, most of our work was inside the electronic analysis part of the vehicle. Not the wild, tree-crushing, roller-coaster rides through the woods part. This was the real intelligence work—monitoring signals, tracking data, doing the kind of Top-Secret operations I had actually been trained to do.

Staff Sergeant Strychland was there. He was new to the unit at that time—Sergeant Shane was out-processing due to an Expiration Term of Service or Permanent Change of Station (ETS or PCS) —and Strychland was replacing him as team leader.

Specialist (Specialist) Burns was there too, along with Staff Sergeant Bleach-Bottle Blond and the rest of our crew. The work was intense, but the setting was incredible.

I remember one morning, stepping out of the vehicle and looking out over the valley below us. The sun was just coming up, and the view was glorious. Majestic even. The kind of thing you see in movies or on postcards but never expect to see in real life.

Mountains rolling away into the distance. Mist in the valleys. The whole world spread out beneath us like a painting. For a moment, I forgot about the cold. Forgot about the work. Forgot about everything except how beautiful this place was to see. I thought: This is why people join the military. For moments like this. For seeing places you'd never get to see otherwise. I did not know then that I would look back on that morning as one of the last truly peaceful moments I would have in the Army.

One day during that training, our company organized a PT event. It was some kind of game—like floor hockey or broomball but played indoors. We were all in our PT uniforms: grey sweatpants and grey t-shirts. The kind of outfit that showed every bit of sweat, and every bit of effort. They set us up in a big room—maybe a gym or a training facility—with shortened brooms and a puck on the ground. The rules were simple: go after the puck, try to score, have fun. And I did have fun.

I was fast. Wiry. Aggressive. I went after that puck like my life depended on it; diving for it; fighting for it; laughing the whole time. At one point, Strychland and I both went for the puck at the same time—our heads in each other's vicinity, not quite a collision, but close. You could tell we were competing against each other.

Everyone was getting into it; Running around, colliding, shouting, and laughing. By the end, we were all sweaty and hot and exhausted in the best way. It was one of those rare moments where I felt like part of the team. Like I belonged. Like I was just another soldier having a good time with my unit. Looking back, I recall playing alongside everyone else, and almost colliding with Strychland. I wonder now if my aggressive play struck him the wrong way; If he watched me dive for that puck and decided I was out of line; If that moment planted something negative in his mind about me. Or maybe he just did not like me for no reason other than I was a Black girl and he was a white man in charge. He never said more than a few words to me anyway. He was the kind of man in authority who grunted instead of talked, who gave orders but never explanations. Looking back, I don't think anything I did would have made a difference. He had already formed an opinion about me. But at that moment, during that game, I was not thinking about any of that. I was just playing. Competing. Laughing. Feeling alive. The Hohenfels training ended, and we headed back to Coffey Barracks. Back to the regular routine. Back to the daily life of a soldier in Germany. I did not know then that things were about to change. That Iraq was about to invade Kuwait. I was just a soldier doing her job, playing games with her unit, looking out over mountain valleys and thinking about how lucky I was to be there. For a little while longer, anyway.

Private Becky had just in-processed to the 511th. I recognized her from Fort Devens. We weren't friends, but I remembered seeing her around. She was also a 98 Juliet. Shortly after Becky arrived, we were both detailed to Nellingen Barracks for a temporary duty (TDY).

The TDY lasted a month or two—long enough that we needed to stay on-site rather than commuting daily. For some reason, they initially put us in a young woman's room temporarily. She was Black, around my age, very pretty. Lower ranking, maybe a Specialist, and I think Army, though I don't quite remember for sure. She may have been in the Air Force.

What I do remember is thinking, even then, about why we were assigned to her room out of all the rooms we could have been assigned to that base.

Was it because she was Black? And nobody gave a damn about her comfort or privacy? Looking back now, I see it so clearly. They put us there because she did not matter to them. Her space, her peace, her right to have her own room—none of it mattered because she was a Black woman soldier and the system had already decided she was disposable. Just like they had decided about me.

Eventually, they moved Becky and me to our own shared room. We were roommates for the rest of the TDY.

We were on different shifts. I was on midnight shift—and Becky was on swings. We were manning a twenty-four-hour operations center doing Top Secret military intelligence work. The strategic intel work that I had always wanted to do.

I was fine with being on different shifts than Becky. I had zero interest in hanging out with her.

Midnight shift destroyed me. I could not sleep during the day. My body wouldn't adjust. I would lie in our shared room during daylight hours, exhausted but wired, and then drag myself to the ops center at night already half-dead from sleep deprivation.

It got so bad I had a breakdown and cursed out my team leader. It was not me—it was the exhaustion, the stress, and the inability to regulate my emotions when I was that tired. I was out of my mind, in the truest sense. I could not control my stress the way I normally could when I wasn't sleep deprived. But here's the thing–my team leader understood—My team leader, a white Staff Sergeant with a loud Black Firebird sports car—got it. We were friends. He knew I was not behaving normally. He knew it was the sleep deprivation talking. I apologized. He accepted my apology and told me to go get some sleep. And we moved on.

That kindness, that understanding, that basic human decency—it stands out in my memory because it was so rare. Most of my leadership did not give me that kind of grace. Most of them were looking for reasons to write me up to prove I was a problem. But this guy? He saw me as a person. Saw me struggling. And helped me instead of punishing me. I wish I could remember his name.

The other thing I can remember about Nellingen is something Becky had done. When we first arrived at that young woman's room, before she came back from wherever she had been, Becky snuck into her refrigerator and ate her yogurt. I did not see her do it. She did it out of my sight; sneaky. If I had seen her take it, I would have said something. But I was not aware until later. We were almost asleep when the young woman came back to her room. She went straight to the refrigerator, then looked around. "Who ate my yogurt?" she asked. Direct. No beating around the bush. I think she saw the empty container in the trash where Becky had thrown it. I was stunned and told her it was not me. I was oblivious to what she was talking about. But Becky... she just sat there looking sheepish.

The young woman did not hold back. "You don't go into someone's refrigerator without permission, and you certainly don't eat their food. This isn't community property. You shouldn't have to tell grown folks this."

She was absolutely right. Becky did not apologize. She just sat there with that "I'm confused" look; like she had been caught but was not actually sorry. She was sickening. No apology. No shame. Just this attitude like she was too important to be bothered by basic rules about not stealing other people's things.

And that's when I really understood who Becky was. Thoughtless, inconsiderate, entitled, and lazy. She wanted people to always give her a break, to make exceptions for her, to treat her like she was special. She acted like she was above everyone and everything.

And the fact that Staff Sergeant Bleach-Bottle Blond let Becky skate by and let her get away with doing nothing while I got every grunt detail—gate guard duty, monitoring the gymnasium, extra CQ shifts and all the manual labor nobody else wanted—it irritated me so deeply. A lot of other people noticed it too. The favoritism was obvious. But I was not in Bleach-Bottle Blond's platoon or squad, so I could ignore it for the most part.

That young woman whose room we stayed in. She was nice to me. Professional. Competent. More than just a pretty face—she was smart, capable, and serious about her work.

She was also young and fun. Before I left Nellingen, she gave me a picture of herself and her friend. They were dressed to the nines in that photo, looking like they were heading out for a night of clubbing. I looked at that picture and thought I so relate.

We were both young Black women trying to make it in a system that did not value us. Both trying to hold onto some version of ourselves outside the uniform. Both trying to have fun and be professional and survive. I still remember her face, even though I can't remember her name. I hope she made it. I hope the military did not break her the way it tried to break me.

The Nellingen TDY ended, and Becky and I went back to our regular units. I saw how Becky operated and how she treated people. Saw her get away with eating someone else's food, skate by on duties, doing the bare minimum—while Staff Sergeant Bleach-Bottle Blond was always on top of me. She checked that I had done my Preventive Maintenance Checks and Services (PMCS) on the vehicles. Verified my motor pool assignments were complete, questioned every task instead of just trusting I had done it. The double standard was obvious. But I was so naïve, I thought that things were just that way. I had no idea it would follow me to the desert and how much worse it could get. But unfortunately, things did get a lot worse.

Chapter 7: The Call of Duty

Iraq invaded Kuwait. August 2, 1990.

At first, it did not feel real. It was news from far away, something happening in a part of the world I had never thought much about. We were in Germany, doing our field exercises, living our lives. War felt abstract.

But by October, the rumors started. We weren't being disbanded and sent back to the States like we were originally told. The Berlin Wall had come down in 1989, and there was talk that the military was downsizing, and closing bases in Germany, and sending soldiers elsewhere under the Base Realignment and Closure program — BRAC.

I was okay with the possibility of being reassigned back to the states. I had my fun in Germany. I ate the food, saw the sights, and lived my adventure. Going back to the States did not sound bad at all. But that's not what happened.

Instead, we were told we would be deploying to the Persian Gulf. Saudi Arabia. Operation Desert Shield. I was devastated. I did not want to go to war. That had never been part of the plan. I had joined the military to pay for college, get out of DC, travel and learn skills and have a better life. War was not supposed to be part of the equation.

But the white boys in my unit–they were thrilled. "Hooah!" they shouted while fist-pumping and talking about how they were finally going to see some "real action"; get some combat experience and prove themselves as soldiers. I thought they were idiots. And I was right.

Once we actually got to the desert and the realities of war set in—the heat, the boredom, the danger, the fear—those same guys were whining and complaining and pouting because Stop Loss took effect, which meant no one could leave when their service was expired and they were deployed. Stop Loss is legal and allowed during wartime.

And it's exactly what happened to all those soldiers who thought they'd be done by their ETS date. They bitched like whiny babies. And I was tickled knowing they were miserable. That's what they get.

But I was miserable too. I did not want to go. I prayed it wouldn't happen. I hoped the orders would change, that the deployment would be canceled, that someone would realize this was a mistake, but it was not a mistake.

ॐ

We shifted from base closure to deployment preparation mode. We had to get all our shots—standard inoculations for infectious diseases. They gave them to us in the buttocks, which was as unpleasant as it sounds. That was just the beginning of a long list of things we had to do to get ready.

We had to pack all our gear, then get it inspected to make sure we had everything: gas mask, rifle, Kevlar helmet, flak jacket, uniforms, and boots. Everything. Oh, and lots of books. If the training was any indication, there was going to be a whole lot of sitting around and waiting.

Our unit, Bravo Company of the 511th Military Intelligence Battalion, was part of VII Corps (7th Corps)—a massive military formation under General Norman Schwarzkopf. VII Corps had multiple divisions, thousands of soldiers, and we were one small Military Intelligence unit among many heading to the Persian Gulf.

At that point, I had no idea what our specific mission would be once we got there. Would we be doing intelligence work at headquarters? Following a combat unit? Where exactly would we end up? All I knew was we were deploying to Saudi Arabia as part of Operation Desert Shield. The rest would be decided once we got there.

Staff Sergeant Strychland took over as my team leader around that time. And that's when things started to change. I did not realize it at the time. Did not put the pieces together. But looking back now, I

48

can see it: the duties started piling up. More than usual. More than made sense.

I was assigned gate guard at the main entrance to Coffey Barracks, every other day, it seemed. I pulled overnight shifts, standing in the cold, checking IDs, and manning the gate for hours. I just thought it was normal. Thought everyone was pulling extra duty because we were preparing for deployment. Thought it was part of the process. I wonder now if Strychland was behind that. I never questioned these things at the time. I was just a soldier following orders. But after seeing him become a sadistic creep in Desert Storm, it was highly possible that Strychland was already targeting me, but I just wasn't aware. Or I could be just a wee bit cynical now that I am 35 years wiser. Life tends to do that to open your eyes that way.

2•

We tried to have a normal Thanksgiving. I don't remember where we went or what we did, but I know we tried. My friends and I, attempting to hold onto some sense of normalcy before everything changed.

My assumption was the deployment meant the base closure was off the table. If VII Corps was vital enough to send to war, surely they wouldn't disband us. Surely Coffey Barracks would remain open when we got back. I was wrong about a lot of things.

But in that moment, packing my bags, I told myself this was temporary. A few months in the desert, then back to Germany and my life here. So I left everything in my room. My boombox. My cuckoo clock. My little hotplate. My clothes, my books, my life. My room stayed my room.

My boyfriend came to visit me from Sinop Turkey right after Thanksgiving. He stayed until December 4th. My old Team Chief, who was moving out of his apartment, let us use his furnished place the whole time Boyfriend was there. But the duties did not stop just because he was visiting. Gate guard. CQ duties or whatever else I was assigned. I would have to leave, and Boyfriend would wait for me. He'd wanted to spend as much time with me as possible before I deployed, but the Army had other plans. I did not question it. I just did what I was told. According to the letter I wrote home on January 5th, I was "so happy" during Boyfriend's visit. I don't remember any of it.

Trauma has a way of erasing the good moments along with the bad. The war would take so much from me—including pieces of my own past. Without these letters, these artifacts I kept, entire chapters of my life would be gone. Lost. Had it not been for that letter, I would never have known it happened at all.

ॐ

After Boyfriend left, the final preparations intensified. More inspections. More packing. More briefings about what to expect in Saudi Arabia. After Thanksgiving 1990, we got our official orders. We were going to Desert Shield.

We deployed in early December 1990. The flight was long. We flew on a commercial airplane—a military charter with just our troops, no other passengers. We spent hours and hours in the air, packed in with all our gear. I tried to sleep but could not. My mind kept racing, thinking about what was ahead. The desert. The heat. The unknown. I looked out the window as Germany disappeared behind us.

Chapter 8: Boots on the Ground

Riyadh, Saudi Arabia. December 1990.

The first thing that hit me when we landed in Saudi Arabia was not the heat, though God knows it was hot enough in December to make me question everything I believed I knew about winter. It was the strangeness.

Everything felt wrong. The air smelled nothing like home. Among the locals, it was like a spice market had exploded—cumin, cardamom, something pungent and unfamiliar—mixed with the unmistakable musk of bodies in the heat. But out in the camps, it was just outdoor dirtiness and mildewed canvas—tents that had been yanked from years of storage and were being aired out for the first time. And the light—back home, it felt like we were sitting in the rear of the arena. Here, we had front row seats to the sun. But front row wasn't better. It was overwhelming, blinding, oppressive. No shade, no clouds, no relief. Just white glare pressing down on everything until your face hurt from squinting. We hadn't just landed in another country. We had landed on another planet.

We filed off the commercial aircraft in our Battle Dress Uniforms, a hundred-plus soldiers squinting in the bright Middle Eastern sun, and I remember thinking: *This is real. This is actually happening.* I had turned nineteen somewhere over the Atlantic, though I don't remember anyone singing or celebrating. Just the hum of the engines and the weight of what we were flying toward. The culture shock was immediate.

Meanwhile, we American women soldiers walked around in our BDUs—baggy, shapeless uniforms that made us look more or less like the men. But we were still women. Women without hijabs. Women with rifles. I could feel the stares.

Our leadership gathered us quickly and gave us the briefing: when we left our staging areas, we were to keep our sleeves down at all times. Long sleeves, no matter how hot it got. It was a sign of respect, they said. Cultural sensitivity. I made a mental note right then: I was not going to venture out much. It was too hot, and I did not want to deal with the added layer of clothing or the added layer of judgment.

Standing there in that first moment, looking out at the vastness of the desert beyond the airport, I felt something I hadn't felt since the day I got on the wrong bus in Germany: *I don't belong here.* Except this time, there was no other bus to catch. No First Sergeant to drag me back to where I belonged. This time, I was exactly where the Army wanted me. And there was no way out.

2❦

The desert itself was both beautiful and terrifying. We drove from the airport to our first staging area in a convoy of military vehicles, and I was just trying hard to process that I was in a war zone. It felt surreal and I was struggling to take it all in. Miles and miles of nothing. Just sand. Beige, endless, empty sand. No trees. No buildings. No streets. No signs of life except for the occasional two-lane highway that seemed to lead nowhere and everywhere at once.

The sky was so big it felt oppressive. Like it was pressing down on us. And the sun, God, the sun, it did not just shine. It *burned*. Even through the window of the vehicle, I could feel it. Then we passed something that took my breath away: the Persian Gulf. The water was the most beautiful blue I had ever seen. Pristine. Clear. Calm. The shoreline was rocky and jagged, dangerous looking, but the water itself looked like glass. Like you could just walk down and take a cool, refreshing swim. Of course, I never would. I did not know how deep it was and I did not trust it. But it was beautiful. That's what the desert was like in those first days: beautiful and untrustworthy.

At first, they put us in a tent city; a temporary base where we would get acclimated before moving deeper into the desert.

The tents were big olive drab (OD) green and made of canvas. They smelled dank and dusty like previously used and put-away-wet military surplus. Inside, there were rows and rows of cots, close together, no privacy. We were assigned bunks and told to set up our gear. I remember unpacking my duffel bag, pulling out my sleeping bag and laying it on the thin cot, and thinking: *This is home now.* There were no walls and no doors and no bathroom down the hall. Just a tent full of soldiers and the constant sound of wind pulling at the canvas. Worse than all that, the food was terrible.

The Saudi's prepared our meals in those first weeks, and I don't know what they were trying to make, but it was not anything I recognized. The eggs were runny and flavorless, like they'd been made from powder and water. The meat was unidentifiable, some kind of mystery protein that tasted like cardboard soaked in grease. I tried a hamburger once. It looked like a hamburger. It smelled vaguely like a hamburger, but it tasted regrettable. Everything was a knockoff version of American food. Like someone had described a cheeseburger over the phone to someone who'd never seen one, and this was their best guess. I ate it anyway. Because the alternative was starving. It made me sorely miss my home and my mother's cooking. I missed food that tasted like love.

There were small pleasures, though. Small moments that made tent city bearable. Some nights, a group of us would gather in someone's tent—usually soldiers from a different squad—and play games by lantern light or whatever generator electricity we could get. We played cards; spades was popular. We also played UNO and I especially loved playing Pictionary.

I don't know why it cracked me up so much, but watching grown soldiers try to draw something recognizable in those conditions—by dim light, under pressure, everyone shouting guesses—was hilarious.

Most of the drawings looked nothing like what the card said. Someone would draw what was supposed to be a horse and it would look like a deformed dog. A telephone would come out looking like a potato with wires. I would be doubled over laughing, tears in my eyes, while everyone else got increasingly frustrated with their terrible artistic skills. Those nights felt almost normal; like we were just a bunch of young people hanging out, not soldiers waiting to go to war. I would not realize then how much I would miss those moments. How the laughter would fade. How the easy camaraderie would shift into something harder, more complicated.

But for those first weeks in tent city, before everything changed, there was Pictionary by lantern light. And I was not miserable.

Then the MREs, Meals Ready to Eat came, and somehow, they were the absolute worst food on planet earth. Each MRE was a brown plastic pouch filled with individually packaged sadness. There was a main entrée (and I use that term loosely), crackers, a tiny bottle of hot sauce, a dessert (usually a rock-hard brownie or a packet of candy), and thank God, a packet of instant cocoa. That cocoa was the only good thing in the entire MRE. It was rich and creamy and tasted like it had been made with real chocolate and real milk, even though all you did was add hot water. It was the one luxury, the one small joy in an otherwise miserable existence. I started hoarding cocoa packets. Trading for them. Begging for them. "Hey, you gonna drink that cocoa?" I would ask anyone who opened their MRE near me. Most people did not care. They'd hand it over, and I would stash it away like gold. The other soldiers started calling me "Cocoa." I didn't mind. It was better than "Bug Eyes."

❧

The worst part of the MREs besides the taste was their names, like Ham and Chicken Loaf. I'll never forget that one. It came in a vacuum sealed square pouch inside the main pouch, and when you opened it, it looked exactly like wet cat food. Grayish-pink mush with chunks of something that might have been meat and a layer of gelatinous goo on top. It was disgusting. But when you're hungry,

54

really hungry, you eat it anyway. You drown it in hot sauce, crumble the crackers into it and close your eyes and pretend it's something else. You survive. That's what we were doing in those early days. Surviving.

ઢ

The duty assignments started almost immediately. They were: KP, Kitchen Police. That meant working in the tent where they prepared and served food. Washing dishes. Hauling trash. Standing in the heat for hours. Water detail. That meant carrying heavy water canisters across the camp to fill up stations. The canisters were five gallons each, and I was maybe 118 pounds soaking wet. I struggled every time. Berm duty. That meant lying on the perimeter in the freezing pre-dawn hours, rifle ready, watching for threats that never came. And my personal favorite: shit-burning detail. But we'll get to that later. I was assigned every grunt job you can imagine. And I did them. Because I had to. But even in those first days, I could feel something shifting. The way people looked at me. The way my team leader, Staff Sergeant Strychland, assigned me tasks. It was not normal. It was not fair. But I did not say anything. I just kept my head down, did my job, and tried not to think about how long I would be stuck here.

ઢ

At night, the desert was a different world. The temperature dropped. The sky opened up. And the stars, God, the stars, they were so bright and so many that it felt like you could reach up and grab them. I would lie on my cot in the tent, listening to the sounds of a hundred soldiers breathing, shifting, snoring, and I would think about home. About my mother who had just become pregnant right before I deployed. After my brother and I were already grown and out of the house, she was starting over. I would think about DC and about the life I had left behind and about the life I expected I would have by the time this was over. And I would pray, quietly and desperately: *Please, God. Get me through this. Keep me safe. Bring me home.*

Looking back, I can see that God was listening. He had plans for me that had nothing to do with my own plans. It never occurred to me that survival was not just about making it out alive, it was about who I would become in the process. The truth was that I was in the desert, and I was alone, and I was scared. And the war hadn't even started yet.

Darnnell and Dawn in Saudi

Chapter 9: January 5th Letter to Mom

King Khalid Military City (KKMC), Saudi Arabia. January 1991.

On January 5, 1991, I wrote my mother a letter. I was sitting in my tent after dinner, flashlight in hand, trying to find words to describe what my life had become. The problem was, I did not really know how to describe it. How do you tell your mother that you're exhausted and lonely and cold and stuck with the worst duties? That I was doing KP on New Year's Day, shit-burning detail and that I had not showered in almost a week? How could I tell her that the days dragged on with nothing but work details; no TV, no phone, no radio—just lying in your sleeping bag by 7:30 p.m. reading by flashlight?

So, I kept it simple, asked about her pregnancy and how fat was she getting; about home, about my brother Glenn, and complained a little about the work. And then, I made a list.

Dear Ma,

I was on duty a lot. What do you do in your spare time, or is all your time spare? You should write me in your spare time. This place reminds me of basic training—when I wanted and needed mail and care packages. I guess I'm hinting that I want one. Actually, I want a lot of them.

I had KP duty on New Year's Day and got filthy. The next day they had me on shit-burning detail, but I went sixty miles away for my first shower in almost a week, so I missed it. It's depressing out here with no TV, no phone, no radio. After dinner I'm usually in bed by 7:30, reading by flashlight in my sleeping bag just to pass the time.

As of January 5, I still haven't gotten your package. I've got a few books, but when I finish them, that's it.

Here's my list of things I really want. I hope you can get these to me—it'll give you something to do when you're not busy.

- *Individually wrapped chocolate candies*
- *Potato chips (Pringles)*
- *Butterscotch candies (hard)*
- *Homemade brownies with walnuts*
- *Articles or magazines (Essence or TIME)*
- *Gates of Paradise by V.C. Andrews*
- *Soap and laundry detergent (liquid)*
- *Book clip flashlight*
- *Towels (bath)*
- *Deodorant, hair moisturizer*
- *Baking soda for my teeth*

This is it for now. If you could do this, I would really appreciate it.

۶❤

Those were the small things that made the difference between just surviving and actually coping. I asked for a book clip flashlight so I could keep reading when the sun went down at 5:30 and I still had plenty of hours to kill before sleep. I needed baking soda because I could not brush my teeth properly without it. The deodorant and hair moisturizer was because I wanted to look presentable and not smell. I wanted to feel human—not just like a body in uniform. Homemade brownies with walnuts was because I needed to taste something from home, something made with love. And books—because when I finished the ones I had, there was nothing left but the sand, the cold, and the waiting.

The desert was not what people imagined—it was freezing in the mornings, with relentless wind, sandflies everywhere, and it even rained. Most days blurred together: KP duty, guard duty, the occasional "shit-burning detail", and a sixty-mile trip just for a shower.

That was my life in early January 1991—no TV, no phone, no radio. In bed by 7:30. reading by flashlight until my batteries died, waiting for mail that took forever to arrive, and begging my mother for soap, deodorant, and a reason to feel human again. This was the reality that doesn't make it into most war stories—the boredom, the isolation, the grinding sameness of each day. The small things like mail and care packages became lifelines.

I asked my mother about my brother Glenn—told her to say hi, that I loved and missed him, and to have him send pictures. I told her to send the packages Priority Mail so they wouldn't take twenty years to get there.

I wrote about my plans—how when I got back to Germany, I wanted to take leave and maybe make it home to the States. Me and Boyfriend had once talked about traveling around Europe, but I suspected I would have different plans by the time this was all over.

I told her I did not hear much about the war itself and asked her to send articles from *The Washington Post*. I wrote: *"I don't hear any news over here as to what's going on, so why don't you mail an article or two from the Post?*

I ended the letter giving her my love:

Love always, your First Daughter DEE

Then I folded it carefully, addressed it in my neat handwriting:

Miss Cheryl
706 Jackson Street N.E.
Washington, D.C. 20017

And on the back flap, I wrote:
Love you! W/B/S — Write back soon.

Please.
Write back soon.

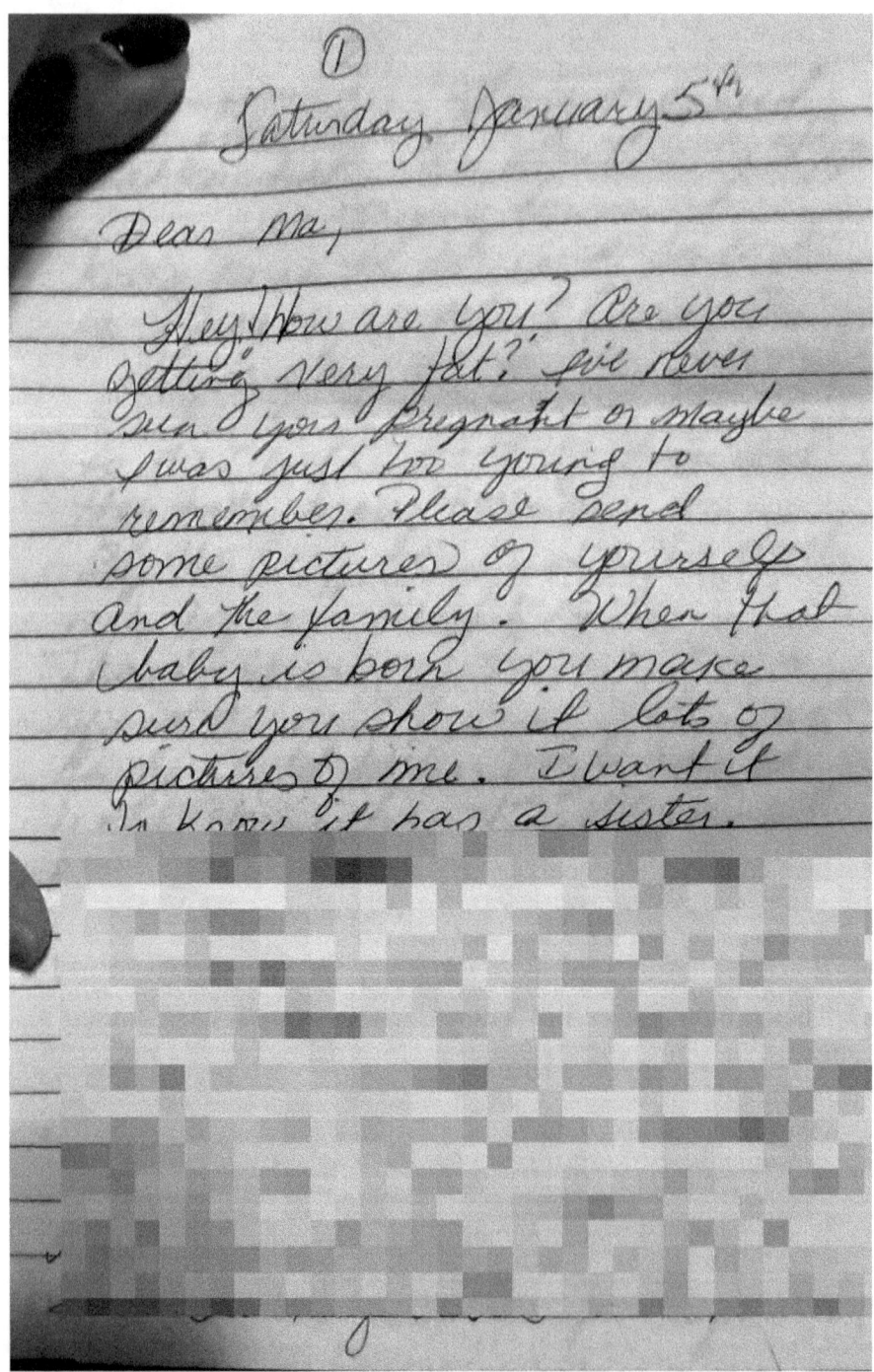

①

Saturday, January 5th,

Dear Ma,

Hey! How are you? Are you getting very fat? I've never seen you pregnant or maybe I was just too young to remember. Please send some pictures of yourself and the family. When that baby is born you make sure you show it lots of pictures of me. I want it to know it has a sister.

2

[illegible — redacted]

By the way he came to visit
me again right after
Thanksgiving. He wanted to
spend as much time with me
as he could before I left. He stayed
until December 4th. I was so
happy. And guess what. My old
Team chief was moving out of
his apartment and was moving
back to the barracks so he
let us use his furnished apartment
apartment the whole time
was here. It was nice of
him, but since we were
preparing for deployment

3

I was on duty alot.

What do you do in your spare time, or is all your time spare. I think you should write me in your sparetime. This atmosphere reminds me of basic training when I wanted and needed mail and Carepackages. I guess I'm hinting around that I want one. But actually I want alot of them. I'll make a list to include in the letter. As of January 5th I still haven't gotten your package.

Its so depressing being out here with no TV or phone no radio. I got some books but when I finished with them that's it. Anyway after work and after dinner I'm usually in Bed by 730pm. I try to read on my cot inside my sleeping bag but the flashlight

Keeps moving. See if you
can find one of these book clip
flashlights. (Nothing fancy)
Today has been pretty slow.
Yesterday I got a shower.
After almost a week without
one. Boy did I feel good. The
water was hot and everything.
OH! I need hair moisturizer.
I wash my hair and condition it
but its very dry in this weather.
Don't let the word desert fool you.
Its cold out here. My butt
freezes every morning. The wind
blows, sand flies and it rained
the other day.
Yesterday they had me on the
shit burning detail. But since
I went to take a shower and it
was 60 miles away I didn't get back
in time to do it. We have to
burn all our feces and crap
thats in the crap buckets.

On New Years day I had
KP and I got so filthy.
What did you and
~~???~~ New Years. I hope you

When I get back to Germany
I'm gonna go on leave and
hopefully I'll make all the way
to the States.
Made plans before all of
this, to travel around Europe.
But I may have other plans
by the time this stuff is over.
I don't hear any news over here
as to whats going on so why
don't you mail an article or
two from the Post.

I really hope this is over soon.
I can't wait to get back.

Well, that's all for now I wish
I could think of something
else but Unfortunately, I can't.
Here is my list of things I
really want. I hope you
can get these to me often.
It will give you something to
do when you're not busy.
Individually wrapped chocolate candies
Potatoe chips (Pringles)
butterscotch candies (hard)
 Home made brownies w/walnuts
articles & Magazines (ESSENCE or TIME)
a book called GATES of PARADISE BY V.C. ANDREWS.
soap and laundry detergent (liquid)
book clip flash light.
towel (s) (bath)
deodorant
hair moisturizer
baking soda for my teeth.
This is it for now. If you would
do this I'd really appreciate it.

7

Would you also tell Glenn
Hi! and I love him & miss
him. Tell him to send me
pictures too. one more thing,
if its not too much money
would you send some of
the care packages Priority so
that it won't take 20 years to
get here.
Well here is the address
again and Thanks very much.
I love you very much and
Glenn.
PFC Adams DARNELL

DESERT SHIELD
B CO 511 MI BN
APO 09748

love always Your Daughter
First
DEE

PFC Adams, Darnnell
DESERT STORM

B Co 511 M I BN
APO 09748

FREE
FREE

Cheryl
706 Jackson Str. NE
Washington, D.C.
20017

Love you!

W/BS

Chapter 10: Stand-To

Remote Desert Temporary Location, Saudi Arabia. January 1991.

The desert cold bit differently than any cold I had known. Back in our Brookland neighborhood in DC, winter meant bundling up and running from the house to the car, and from the car to the building. It meant layering up—scarf, long johns, thick socks, boots—hood pulled over your hat, arms crossed tight with your mittened hands buried in your armpits, waiting for that bus.

But in Saudi Arabia, in the earliest hours before dawn, what we called "o'dark-thirty", the cold was sharp and mean. Lying prone on that cold, hard sand for one to two hours is what made it unbearable. The cold came up from the ground, seeped through your uniform, settled into your bones. And every single morning, we had to lie in the sand. Stand-to is what it's called it, though there was no standing involved. Just lying prone on a small incline, a berm, really, nothing more than a rise in the endless sand, with our weapons aimed forward into the darkness, six feet apart from the next soldier, waiting for an enemy that might try to ambush our camp. We did these one or two hours, every morning, in temperatures that dropped to thirty degrees, or below.

The first time I had to do it, I believed I had prepared myself. I wore every layer I had, brown tee shirt, BDU top, field jacket. I pulled my patrol cap low. I tried to make myself small against the cold. It did not matter. The cold found me anyway.

I lay there in the prone position, M16 aimed at nothing, watching the horizon for movement I could not see in the dark. My breath came out in small clouds. My fingers, wrapped around the rifle, started to go numb despite my gloves. My nose started running—I had stuffed toilet paper in my sleeve for emergencies like this, wiping quietly, trying not to move too much.

Around me, I could see the other soldiers. Mostly men. Mostly silent except for the occasional cough or the rustle of someone shifting position. No one talked during stand-to. You just lay there and waited for the sky to lighten, for the danger window to pass, and for permission to get up and feel your legs again.

That first morning, I made it through. Barely. But then it became every morning. And then it became routine. And then it became one more thing my body had to endure in a place that seemed designed to break it down piece by piece.

॰

The duties started piling up after we had been in-country for a few weeks. At first, I thought maybe it was normal. Maybe everyone was getting assigned the same rotation of terrible tasks: kitchen police, water detail, berm duty, guard shifts. But then, I started noticing, watching, and paying attention to who was doing what.

Specialist Burns was not carrying water jugs across camp every day. Specialist Fox was not pulling KP duty every rotation. Specialist Quest was not getting smoked for minor infractions that weren't even infractions. It was just me.

"Adams!" Strychland barked one afternoon when we had just finished a long convoy movement and were setting up camp in a new location. "Water detail. Two canisters. Don't come back until they're full." I looked at him, then at the rest of the team. Burns was lighting a cigarette. Fox was already reclining on his cot. Quest was checking his weapon. None of them were told to haul water. "Roger, Sergeant," I said, because what else could I say?

The water canisters were heavy; each one weighed forty, maybe fifty pounds when full. I was 118 pounds soaking wet, with arms like sticks and no real upper body strength. And just like I had barely been able to do one push-up when I got to basic training, but I picked up those canisters and started walking.

The potable water station was about an eighth of a mile away. It was not far, but far enough when you're carrying weight in soft sand, in the heat, with the sun beating down on you and no shade anywhere. I made it to the station, filled both canisters, took a breath, then I picked them up and started back to the tent.

My shoulders screamed. My arms shook. The handles dug into my palms even through my gloves. I had to stop twice just to put them down and shake out my hands. When I finally got back to the tent, Strychland barely looked at me. "Set them down over there," he said, gesturing vaguely. "We're gonna need more tomorrow." I wanted to say something. I wanted to ask why it was always me and why the men on our team, bigger, stronger, more capable, weren't being assigned the same tasks. I already knew the answer he did it, because he could.

≥❧

I guess I wasn't picking up what Strychland was putting down. My younger self did not see it, but my more mature eyes recognized it now—he was provoking me on purpose, waiting for me to give him a reaction he could use against me. When the water canisters did not do it, he tried something else. One morning, after we had moved to a new location away from King Khalid Military City, Strychland called me out in front of the team. He didn't mind publicly humiliating me. "Adams, you're responsible for the tripod," he said, nodding toward the heavy metal stand used to mount the .50 caliber machine gun. "You carry it everywhere. Chow, latrine, wherever you go. You don't put it down unless I say so." So, on top of being the water bearer, I was now the tripod carrier as well. I stared at him and said, "Sergeant, that's—" he cut me off. "Did I stutter, Private?" "No, Sergeant." I said and Strychland responded, "Then pick it up."

The tripod weighed at least twenty pounds. It was awkward, unwieldy, with sharp edges that dug into your shoulder or your side no matter how you carried it. And I had to take it everywhere, in addition to my M16, my gear, and my rucksack. I looked like a pack mule.

The other soldiers noticed. Some of them looked away, uncomfortable, but silent. Burns smirked. Fox made a joke I did not catch but heard them laugh at it. Only Quest said anything.

70

"Sergeant, you want me to carry that?" he asked quietly, nodding toward the tripod. Strychland's head snapped toward him. "Did I ask you, Quest?" "No, Sergeant, but—" "Then stay in your lane."

Quest's jaw tightened, but he did not push it. He shot me a look, something like sympathy, maybe pity, but he did not say anything else. I picked up the tripod and slung it over my shoulder. And I carried it, for weeks.

I took it to the chow tent, even though we weren't eating in a chow hall anymore, just grabbing MREs, to the latrine, (which meant walking across camp in the dark with a rifle, a tripod, and a flashlight, praying I did not trip over a tent line) to formation, to guard duty, to every single place I went.

Looking back, I think Strychland felt like he was taking candy from a baby. He was watching me like a hawk, waiting for the exact moment I would slip up.

One day, I needed to use the latrine. I asked my roommate from Coffey Barracks—the one who was always off with her fiancé—if she would watch my gear (Rucksack, weapon, and tripod) while I stepped away. She agreed. I took too long, and when I came back, she was gone. My gear sat there unattended. But Strychland was waiting.

He put me on thirty-day restriction with no town privileges, and no phone calls home for thirty days. While other soldiers went off site to call their families, I had to stay behind. One more punishment. One more way to isolate me.

I carried that tripod until my shoulder was bruised, until the skin under my uniform was raw. I wanted to drop it in the sand, walk away and let them court-martial me because at least then I would be done. I didn't because that's what Strychland wanted me to do. He wanted me to break, to fail. He was looking for a reason to write me up, to punish me further, to prove whatever point he thought he was making. So, I carried it. And I quietly hated him.

It was Quest who finally helped me without asking. One afternoon, after a particularly brutal stretch of duty rotations, I was walking back from the latrine with my rifle slung over one shoulder and the tripod over the other. My arms were shaking. My legs felt like they might give out. I was so tired, I could not think straight. Quest appeared beside me, walking in the same direction. "Here," he said quietly, reaching for the tripod. "I got it," I said automatically, because I knew what would happen if Strychland saw someone else carrying it. "I know you do," Quest said. "But I got it now." He took the tripod off my shoulder before I could argue and slung it over his own like it weighed nothing. I wanted to cry. Not because I was weak, I knew I was not weak, not really, not where it mattered, but because someone had finally seen me. Someone had finally acknowledged that what was happening was not right. "Thanks," I managed to say. He just nodded and kept walking.

After that, whenever Strychland was not looking, Quest would help. He'd carry the water jugs with me. He'd take the tripod for part of the walk. He'd cover for me when I needed an extra minute to catch my breath. He never made a big deal out of it. Never asked for anything in return. He just helped. And that kindness, small, quiet, consistent, was the only thing that kept me from completely breaking.

One night, about a month into our deployment, I was sitting outside the tent after another long day. The sun had just set, and the temperature was starting to drop. My whole body ached. My hands were blistered from carrying the water canisters. My shoulder had a permanent bruise from the tripod. Quest came and sat down beside me. He did not say anything at first, just pulled out his foot soak, a small basin filled with water and bleach, and started unlacing his boots. "You good?" he asked after a while. I almost laughed. Was I good? No. I was exhausted, isolated, and being systematically punished by a man who had all the power and I had none. But I did not say that. "Yeah," I said. "I'm good." He glanced at me, and I could tell he did not believe me.

"He's an asshole," Quest said quietly, soaking his feet. "Strychland. You know that, right? It's not you." I looked at him, surprised. "I know you think it's something you did," he continued, staring at his feet in the basin. "But it's not. He's just an asshole. And he picks on people he thinks won't fight back."

"I'm not gonna fight back," I said bitterly. "What am I supposed to do? He's a Staff Sergeant. I'm a private. He can make my life hell and there's nothing I can do about it." "You're already doing it," Quest said. "You're surviving him. That's the fight."

I did not know what to say to that. We sat there in silence for a while, him soaking his feet, me trying not to cry from exhaustion and frustration and the crushing weight of knowing this was my life now, and there was no way out.

"Thanks," I finally said. "For helping with the water and the tripod and... everything." He shrugged. "You'd do the same for me."

I was not sure I would have. I was not sure I had the strength to help anyone when I could barely help myself. But I nodded anyway. Because in that moment, Sonny Quest was the only person in the entire desert who made me feel like I was not alone. And I held onto that feeling.

Chapter 11: The PB Pills

January 1991. Staging area, Saudi Arabia.

They handed them out during formation one morning in late January, about a month after we had arrived in Saudi Arabia. Little white pills in blister packs. Twenty-one tablets per pack, with the days of the week printed on the foil backing, so you wouldn't lose track.

"Pyridostigmine bromide," the medic announced, holding up a pack like he was showing us a prize. "PB pills. Nerve agent pretreatment. You take one every eight hours. No exceptions." I looked down at the blister pack they'd given me, turning it over in my hands.

"What's it for?" someone asked. "Protection," the medic said. "In case Saddam uses chemical weapons, specifically soman, a nerve agent. These pills won't stop it, but they'll give you a better chance of surviving if you're exposed." A better chance. Not a guarantee. Not protection. Just a better chance.

"What are the side effects?" another soldier called out. The medic hesitated. "Some people report nausea, stomach cramps, increased salivation. Maybe some fatigue. But nothing serious. And it's better than dying from nerve gas, right?" A few nervous laughs rippled through the formation. I did not laugh. I looked at the pills in my hand and felt a knot forming in my stomach.

"What if you don't want to take them?" I asked, my voice smaller than I meant it to be. The medic's expression hardened. "It's not optional, Private. This is a direct order. You take the pills, or you're in violation of a lawful command. Understood?" "Understood, Sergeant."

But I did not understand. I did not understand how they could force us to take medication we did not want. How they could tell us it was

74

safe when they clearly did not know. How they could make it an order, as if our bodies were military property and we had no say.

I looked around at the other soldiers. Most of them were already popping the pills out of the blister pack, dry-swallowing them, moving on. No one else seemed worried. But I was worried. This was my body and I had no control over what they were putting in it. I stood there holding that blister pack, feeling the weight of it in my hand, and I prayed. *God, please. If I have to take these, please protect me. Please don't let them hurt me. Please keep me safe.* Then I pushed the first pill through the foil and swallowed it. Thankfully, I had no side effects. But some soldiers did. Some complained of nausea, or a stomachache. And that scared me.

Every eight hours, like clockwork, I had to take another pill. Morning, afternoon, night. Over and over. Three times a day for weeks. Some soldiers did not seem to notice any effects. Others complained about the sweating, the nausea, the weird metallic taste in their mouths, but no one refused because refusal meant disciplinary action, being written up and proving you weren't a team player and weren't willing to follow orders, weren't really a soldier. So we all swallowed the pills. And we hoped they wouldn't kill us before the enemy did.

❧

The mustard gas alert came without warning. I was walking back from the shower tent one evening, one of the rare times I had managed to get clean water and a few minutes to wash the sand and sweat off my body. I had changed into my PT uniform, brown tee shirt and BDU pants, and left my gas mask back in the tent. Stupid. I knew better. We were supposed to have our masks on us at all times, attached to our thighs or slung over our shoulders. I just wanted a few minutes of feeling normal, of not being weighed down by gear, of not being on high alert even though I knew better. Then the siren went off. A piercing, wailing sound that cut through the camp like a knife. My blood went cold. I knew that sound. We had drilled for it back in Germany, back when deployment was just a theoretical possibility and chemical warfare was something we read about in field manuals. But this was not a drill. This was real.

I started running. The tent was about a hundred yards away, close, but not close enough. My lungs burned. My legs pumped. My heart slammed against my ribs so hard it felt like it might break through. Around me, other soldiers were scrambling, shouting, pulling on their masks as they ran. Some were already suited up in their MOPP gear, Mission Oriented Protective Posture, full chemical suits that made you look like an astronaut and feel like you were suffocating. I did not have time for that. I just needed my mask. I burst into the tent, nearly tripping over a tent line, and grabbed my mask from where I had left it on my cot. My hands were shaking so badly I could barely get it out of the bag. *Come on, come on, come on.*

I yanked the mask over my head, pressed it against my face, and took a deep breath to create the suction seal. The rubber pressed tight against my skin. The straps dug into the back of my head. The lenses fogged slightly as I exhaled. But I could breathe. I sank down onto my cot, mask on, heart racing, and waited. Outside, the camp was chaos. Soldiers running. Voices shouting. The siren still wailing. And then, after what felt like hours but was probably only minutes, the all-clear sounded. False alarm. Or maybe not false, maybe the gas had been detected but hadn't reached us. Maybe the wind had shifted. Maybe we had gotten lucky.

The air war had been going on for over a week by then. Every night, we would hear the distant rumble of bombs falling on Iraqi positions. We saw flashes of light from distant explosions on the horizon. We knew that somewhere "over there," the war was happening, but it felt distant and abstract. Like watching a storm from far away. The Scud alerts made it real. Iraqi missiles capable of carrying chemical weapons. When the alarm sounded, we had seconds to mask up. Iraq was firing back. And we were in range. That's when I understood: the war was not just "over there" anymore. It was here.

I pulled my mask off my face, inhaling the cool fresh air, and my hands were still shaking. The adrenaline flooded my system. I kept thinking about the PB pills I had been taking, the chemicals I had been breathing, the toxins I could not avoid no matter how careful I was.

And I prayed again. *Please, God. Please keep me safe.* I did not know if He was listening. But I had to believe He was. Because if He was not, then I was completely alone out here.

❧

After the gas alert, I became obsessive about my mask. I wore it on my thigh at all times, even when I was just walking from one tent to another. I checked it compulsively, making sure the seal was good, the filter was clean, and the straps were tight. I could not let that happen again. Could not risk being caught without it. Next time, it might not be a false alarm, the gas might be real, and I might not make it.

The other soldiers noticed, but no one said anything. We were all paranoid now. All hyper-aware of the invisible threats around us. Saddam had used chemical weapons before, on the Kurds, and on the Iranians. We saw the pictures in training. Where bodies were lying in the streets, twisted and still, and whole villages wiped out.

It could have happen to us. And all we had between us and that fate was a rubber mask and a prayer.

Chapter 12: Guard Duty Consequences

It was late January when everything changed. I did not know it at the time. Did not feel the shift in the air or sense the weight of what was about to happen. I was just trying to get through another night, another duty rotation, another twenty-four hours in the desert. But that night, that one night, set everything in motion.

Guard duty. They assigned me and Sonny Quest to the overnight shift together. Twelve hours in a Humvee, parked at the edge of camp, watching for threats that probably wouldn't come but might. Eyes on the horizon. Radio at the ready. Two soldiers, alone in the dark. It should have been just another shift. But it was not.

꙳

The Humvee was cold when we first got in, the metal interior holding onto the chill of the desert night. Sonny took the driver's seat out of habit, even though we weren't going anywhere. I sat in the passenger seat, rifle between my knees, scanning the darkness through the windshield.

For the first hour, we barely talked. That was normal for guard duty. You stayed alert, quiet, and focused. You did not want to be the soldier who got caught slacking, who missed something important because you were chatting instead of watching. But then the silence started to feel heavier. Not uncomfortable, exactly. Just... significant.

"You good?" Sonny finally asked, his voice low. I glanced at him. "Yeah. Why?" He shrugged. "You've been quiet lately. More than usual." I almost laughed. I had been quiet because I'm being targeted by my team leader and was tormented by unending duties. And had no idea how I was going to survive this deployment, much less the rest of my military career. But I could not say any of that. "Just tired," I said. "Yeah," he said. "Me too."

Another stretch of silence. Then Sonny said, "You ever think about after?" "After what?" "After this. The war. Going home." I looked out at the dark desert, at the nothingness stretching in every direction. "All the time," I admitted. "What do you think you'll do?" I didn't know how to answer that. I did not even know if I would make it home. Did not know what my life would look like on the other side of this. "I don't know," I said quietly. "Maybe go to college. If I can." "You should," Sonny said. "You're smart. You'd be good at it."

The compliment caught me off guard. No one had said anything like that to me in months. Everyone here treated me like I was a problem, a burden, someone who did not belong. But Sonny did not look at me like that. He never had. "What about you?" I asked, deflecting. He smiled a little. "Go back to Connecticut. See my family. Maybe settle down somewhere quiet. Have a normal life, you know?" "Yeah," I said. "I know."

We fell into another silence, but this one felt different. Softer. And then Sonny said, "You deserve better than this, you know." I looked at him. "What?" "This," he said, gesturing vaguely at the desert, at the camp, at everything. "The way Strychland treats you. The way everyone just... lets it happen. You deserve better." My throat tightened. I did not know what to say. Did not know how to respond to someone seeing me, really seeing me, when I had spent so long feeling invisible. "Thanks," I managed. He nodded. "I mean it."

It was sometime after midnight when the conversation shifted again. We had been talking about home, about family, about the small things we missed that you don't think about until they're gone. Hot showers. Real food. Sleeping in a bed that did not smell like sand.

And then Sonny said, "You ever think about having kids?" My stomach dropped. I don't know why he asked. Maybe it was just the randomness of late-night conversation. Maybe he'd been thinking about it. But the question hit me like a punch.

"Yeah," I said carefully. "I do." "Me too," he said. "I think I would be a good dad." "You would," I said. And I meant it. Sonny was kind. Patient. The kind of person who helped without being asked, who noticed when someone was struggling and did something about it. He'd be a good dad.

He looked at me then, really looked at me, and something in his expression shifted. "What if we did?" he said quietly. "Did what?" "Had a kid. Together." I stared at him. "Not now," he said quickly. "I mean, after. After all this. When we get home. What if we got married? Had a family?"

My heart was pounding. This was crazy. We were in the middle of a war. We barely knew each other outside of this deployment. We were kids ourselves, I was nineteen, he was maybe twenty-one. But the way he was looking at me, the way he said it, like it was a real possibility, like he'd been thinking about it, made me want to believe it. Made me want to believe in a future where I was not alone.

"You're serious," I said. "Yeah," he said. "I am."

I did not know what to say. I remembered my boyfriend, who I had left behind in Turkey. About whether we were even still in a relationship. I had not received one letter from him. It felt as if the desert changed everything. And Boyfriend was not here. And Sonny was.

"Okay," I said finally. "Okay?" "Yeah," I said. "Let's do it. Let's... have a baby." The words felt surreal coming out of my mouth. But also right, somehow. Like maybe this was the answer. Maybe this was how I survived. Sonny smiled, really smiled, and reached for my hand. "We're gonna be okay," he said. "After this. We're gonna have a good life." And in that moment, I believed him.

❧

I don't know who moved first. Maybe it was me. Maybe it was him. Maybe it was both of us at the same time, pulled together by loneliness and fear and the desperate need to feel something other than despair. But suddenly we were kissing, and then we were climbing into the back of the Humvee, into the open cargo area where there was just enough room for two people if they did not mind being close. We did not mind.

We fumbled with uniforms, with buckles and buttons and layers. It was awkward and hurried and nothing like the movies. But it was also tender. Careful. Like we both knew this mattered, even if we did not fully understand how yet. And when it was over, we lay there in the back of the Humvee, tangled together, breathless and quiet. "Are you okay?" he asked. "Yeah," I said. "I'm okay." He kissed my forehead. "Good." We stayed like that for a while, listening to the desert night, to the sound of our own breathing. And then we saw headlights.

෨

"Shit," Sonny said, sitting up fast. A vehicle was approaching, coming from the direction of camp, heading toward our position. We scrambled to get dressed, pulling on uniforms, tucking in shirts, trying to look like we had been doing what we were supposed to be doing this whole time.

The vehicle slowed as it got closer, then stopped a few yards away. Sonny jumped out of the Humvee, waving his arms. "Wait!" he shouted. "Hold up!" I watched from the back as he ran toward the vehicle, flagging it down.

My heart was pounding again, but for a different reason now. If that was someone checking on us, if they'd seen what we were doing, we were done. Court-martialed. Dishonorably discharged. Sent home in disgrace. I prayed it was not.

Sonny reached the vehicle, talked to whoever was inside for a minute, then jogged back. "Who was it?" I asked, climbing back into the front seat like nothing had happened. Sonny's face was pale. "First Sergeant," he said. "She was transporting a new soldier into our unit."

"Did she say anything about why we did not stop her?" I asked. "Nope, she was on a mission," he said. "Good," I said. "Who's the new soldier?" "Don't know, some dude," Sonny said indifferently. "Cool," I said. Relieved that the first sergeant hadn't caught us with our pants literally down. I let out a breath I did not know I was holding. We had gotten lucky. But luck like that doesn't last.

≈

The rest of the shift passed in afterglow silence. We did not talk about what had just happened. Did not acknowledge it. Just sat there, watching the dark, pretending everything was normal. But it was not normal. And I knew, even then, that nothing would be normal again.

When our relief showed up at dawn, we handed over the radio and the keys and walked back to camp together. Sonny reached for my hand as we walked, and I let him.

"That was real, right?" he asked quietly. "What we said? About having a baby, about being together?" I looked at him, this kind, sweet man who'd helped me carry water and listened to me when no one else would, and I nodded. "Yeah," I said. "It was real." He squeezed my hand, and I tried to believe that we would actually get that future. That this wouldn't all fall apart. That maybe, just maybe, something good could come from this war.

But even then, in the back of my mind, I knew. I knew that life doesn't work that way. That promises made in the dark don't always survive the light of day. That what we had just done—impulsive, reckless, born of loneliness and need—would have consequences I could not predict. I knew. I just did not want to admit it yet.

Chapter 13: Boyfriend Arrives

Dawn found me the next morning. I was sitting outside our tent, trying to process everything that had happened the night before. The guard duty with Sonny. What we had done. What we had promised each other. The vehicle that had driven by. Sonny running after it.

I was still in that hazy, uncertain space between hope and dread when Dawn came rushing towards me, her eyes wide with excitement. "Guess what?" she said, practically bouncing. I looked up at her. "What?" "Boyfriend is HERE!" I blinked. "Boyfriend who?" She laughed. "Your Boyfriend! The First Sergeant brought him in the middle of the night!" My world tilted. My Boyfriend. Here. In Saudi Arabia. No. "What?" I said again, stupidly. "Yeah! He's here! He came all this way!" Dawn was grinning like this was the best news in the world. And maybe for anyone else, it would have been. But for me? All I could think was: The middle of the night. The First Sergeant. The vehicle that drove by. *Oh God. Oh God, no.*

༄

I don't remember how long I sat there, frozen, trying to make sense of the timeline. Sonny and I in the Humvee. The vehicle driving past. Sonny running after it, catching it. Coming back and saying, "It was the First Sergeant." And now Dawn telling me my boyfriend had arrived with the First Sergeant in the middle of the night. The middle of the night. The same night. The exact same time. He was in that vehicle. My boyfriend was in the vehicle that drove past us right after Sonny and I— I was going to be sick.

I don't know how long it took before he found me. Maybe minutes. Maybe hours. Time had stopped making sense. But suddenly he was there, standing in front of me, and he was smiling. That big, genuine, happy-to-see-you smile that I used to love. "Darnnell!" he said, and he pulled me into a bear hug. I stood there, stiff, unable to move.

"I can't believe I'm finally here," he said, pulling back to look at me. "I've been trying to get here for weeks. I had to pull so many strings, talk to so many people, but I did it. I volunteered to deploy. I told them I wanted to be stationed where you were." He was so happy. So proud of himself. So sure this was a good surprise.

"You... you came because of me?" I managed to say. "Of course I did!" He was still grinning. "I know things were hard between us before you left Germany, but I could not stand being apart from you. I had to see you. I had to be here with you." My stomach twisted.

"When did you get here?" I asked, even though I already knew the answer. "Last night. Well, technically early this morning. The First Sergeant brought me in. It took forever to get through all the checkpoints and find the right unit, but we finally made it around—I don't know, maybe 2 or 3 AM?"

There it was. Confirmation. He was in that vehicle. The one that drove past the Humvee where Sonny and I had just— "Oh my god!" I said, and my voice came out flat. His smile faltered. "What's wrong?" I looked at him. This man who'd traveled across a war zone to be with me. Who'd volunteered for deployment. Who'd done everything he could to find me. And I had just slept with someone else hours before he arrived. "I need to tell you something," I said.

I told him. Not everything. Not every detail. But enough. Enough for him to understand what had happened. Enough for the smile to die completely on his face. "Last night?" he said. His voice was very quiet. I nodded. "You're telling me that while I was in that vehicle, driving around trying to find you, trying to surprise you, you were—"

"I did not know you were coming," I said, and I hated how defensive I sounded. "You did not write. I hadn't heard from you in weeks. I thought—" "You thought what?" Now his voice was rising. "That I had just forgotten about you? That I did not care?"

Chapter 13: Boyfriend Arrives

"I did not know what to think!" I shot back. "You were in Turkey. I was here. We hadn't talked in—" "Because I was trying to get HERE!" he shouted. "I was in transit! I was pulling every string I could to get assigned to your location! I could not write because I was moving!"

People were starting to look. I lowered my voice. "Babe, I'm sorry. I'm so sorry. I did not know—" "Who is he?" I hesitated. "Who. Is. He." "His name is Sonny. He's on my team. He's been... he's been helping me. With everything. Staff Sergeant Strychland has been making my life hell, and Sonny's the only one who—" "So you slept with him." It was not a question. "It was not like that," I said weakly. "Then what was it like, Darnnell? Explain it to me. Make it make sense."

But I could not. Because it did not make sense. None of it made sense. I had made plans with Sonny. Talked about having a baby with him, about being together after the war. And hours later, the man I had been with before—the man I had loved, maybe still loved—was standing in front of me, and I had just destroyed him. "I can't," I whispered.

He stared at me for a long moment. Then he shook his head. "I came all this way for you," he said. "I gave up my assignment in Turkey. I volunteered for a war zone. Because I wanted to be with you. Because I thought we had something worth fighting for." "Babe"— "And you could not even wait one night."

His words hit like a slap, because he was right. If I had waited. If I had just waited one more night. One more day. But I hadn't. And now everything was broken. "I'm sorry," I said again, and I meant it. But sorry was not enough. It was never going to be enough.

He looked at me one more time, and I saw everything in his face— the hurt, the betrayal, the death of whatever we had. Then he turned and walked away. "Wait," I called after him. "Please, just—" He stopped. Turned back to face me.

"I did not plan it," I said, my voice shaking. "I thought I would never see you again. I thought we were done." "So, you just moved on? Just like that?" "It was not like that—" "Then what was it like, Darnnell? Explain it to me. Make me understand how you could do this."

I did not have an explanation. Not one that would make it okay. "I'm sorry," I whispered. My boyfriend laughed, bitter, hollow. "You're sorry." "I am." "Yeah, well, sorry doesn't fix this." He turned and walked away, and I let him go. Because what else could I do?

ॐ

I found Sonny later that evening, sitting outside our tent cleaning his rifle. The sun was starting to set, turning the sand copper and gold. When he saw me approaching, something in his face shifted—he already knew. "So," he said, not looking up from his work. "That was him."

I sat down on the sand a few feet away, not close enough to touch. "Yeah. That was Boyfriend." He nodded slowly, his jaw working. "The vehicle we almost did not stop." "Yeah." The silence stretched between us like the desert itself—vast and uncomfortable.

"I did not know he was coming," I said quietly. "I swear, Sonny. I thought... I thought we were done. That distance and deployment would've—" "Ended it," he finished. His hands stilled on the rifle. "That's what you thought." "Yes."

He finally looked at me, and the hurt in his eyes made my chest tight. "But it did not." "No." I picked at the sand between my boots. "He volunteered to deploy. Got himself reassigned to this sector somehow. Took him weeks to get here, which is why I hadn't heard from him. He was in transit the whole time."

"Jesus." Sonny let out a breath and set down his rifle. "He came to a war zone. For you." The weight of that hung in the air between us.

Chapter 13: Boyfriend Arrives

"I told him," I said. "About us. Not everything, but... enough."
"How'd that go?" "About how you'd expect." My voice cracked a little. "He's hurt. Angry. Confused."

Sonny rubbed his face with both hands. "And what did you tell him about... what we talked about? About the plans we made?"

I looked at him then, really looked at him—this man who'd shown me kindness when everyone else had shown me cruelty. Who'd carried water jugs without being asked. Who'd helped me dig foxholes when Strychland was working me to the bone. Who'd made me believe, even for a moment, that someone could see me as worthy of protection. "I told him I made a mistake," I whispered. "That I was confused and lonely and—"

"A mistake." His voice went flat. "Sonny, I did not mean—" "No, I get it." He stood up, brushing sand off his uniform. "You've got a boyfriend who crossed the world to be with you. And I'm just... what? The guy who was convenient when you were lonely?"

"That's not fair." "Fair?" He laughed, but there was no humor in it. "Darnnell, we made plans. We talked about getting married, having a life together when we got home. And now, less than twenty-four hours later, I'm a mistake."

"I never said you were a mistake," I said, standing too. "What happened between us—it mattered. You matter." "Just not enough." The words hit like a slap. "I can't do this right now," I said, my voice barely above a whisper. "I can't figure out what's real and what I wanted to be real while both of you are here, watching me, waiting for answers I don't have." Sonny's jaw tightened. "So you're choosing him."

"I'm not—it's not that simple." "Sounds pretty simple to me." His voice was harder now, the hurt turning to something sharper. "He shows up, and suddenly everything we talked about—all those plans—they just disappear?"

"Sonny, please—" "Did you mean any of it? Or was I just... what? Someone to pass the time with until your real boyfriend came back?" "That's not fair." "What's not fair is you acting like we have to figure this out together when you've already made your choice."

"I haven't made a choice! I'm just trying not to hurt anyone else—" "By hurting me instead?" He shook his head. "You think stepping back, giving yourself space, whatever you want to call it—you think that doesn't tell me exactly where I stand?" The words hung between us, cutting and true.

"I'm nineteen years old in a war zone," I said, hearing the desperation creep into my voice. "Everyone's already watching me like I'm some kind of spectacle. I've got Strychland giving me every grunt detail he can think of. I've got two men I care about who both feel betrayed. I don't know what the right answer is, Sonny. I'm just trying to survive this."

"By pushing me away?" "By not making things worse!" "For who?" His voice dropped, but the edge stayed sharp. "Not worse for you, you mean. Not worse for him. But me? I'm just supposed to what—wait around while you figure out if I was ever actually an option?"

I felt tears burning behind my eyes. "You were never just an option. What we had—it mattered." "But not enough." He said it flat, like a door closing. "Sonny—" "No, I get it." He picked up his rifle, slinging it over his shoulder with more force than necessary. "You want space? You got it. But don't pretend you're doing this for my benefit. You're choosing him, Darnnell. Just be honest about it."

"That's not what I'm doing—" "Then what are you doing?" He turned back to face me fully, and the rawness in his expression made my chest ache. "Because from where I'm standing, you're telling me that everything we said to each other last night doesn't count now that he's here. That I don't count."

Chapter 13: Boyfriend Arrives

"You do count," I whispered. "That's why I can't—I can't keep doing this to you. To either of you." "So you're taking yourself out of the equation. Making the choice by not making one." He nodded slowly, like something was settling into place. "That's real convenient, Darnnell. You get to be the victim in all this, and I get to be the guy who was not good enough to fight for."

The words hit like a physical blow. "I never said that." "You did not have to." He started walking away, and I felt panic rise in my throat. "Sonny, wait—" He stopped but did not turn around. His shoulders were rigid, his whole body tense.

"I meant what I said last night," he said quietly. "Every word of it. I would've built a life with you. Married you. Given you everything I had." He looked back over his shoulder, and the hurt in his eyes nearly undid me. "But I'm not going to beg you to choose me over someone else. I'm worth more than that. And you should be too."

He walked away then, and I did not try to stop him. I stood there alone as the sun sank lower, turning the sand to copper and shadow. Sonny was right. I was making a choice by refusing to choose. And maybe that made me a coward. Maybe it meant I was taking the easy way out, but nothing about this felt easy.

Boyfriend, who'd crossed the world thinking we still had something worth saving. Sonny, who'd made me believe I could have something different. Something better than what I had known. And me, caught between two futures I had promised to two different men, knowing I could not deliver on either without destroying someone.

I did not know then that in a few days, none of it would matter. That the worst thing this desert would take from me was not a choice between two men—it was the illusion that I had ever had any control at all.

I turned and walked back to my tent, already feeling the weight of something I could not yet name. Maybe he was right. Maybe I was choosing to walk away from both of them. But it was the only choice

I could make that did not destroy someone else. Even if it destroyed me instead.

❧

After that, I was alone. My boyfriend wouldn't talk to me. Sonny wouldn't look at me. Dawn tried to be supportive, but even she did not know what to say. I became the story people told. The girl who'd slept with another guy in the desert. The girl who caused drama. And I let them say it, because what did it matter? They were going to think what they wanted to think. And I was too tired to fight it anymore. So, I stopped talking to people. Stopped trying to explain. Stopped trying to defend myself. I just kept my head down, did my duties, and waited for this deployment to be over. Waited for the day I could go home and leave all of this behind.

But even then, I knew. You can leave a place. But you can't leave what happened there. It follows you. Forever.

Chapter 14: The Night Everything Changed

Early February 1991. Remote Saudi Arabian Desert heading for Iraq.

The convoy stretched across the desert like a segmented serpent, dusty and endless. Humvees, the MSQ1O3 Charlie, deuce-and-a-halves, APCs (Armored Personnel Carriers) and 2nd ACR mechanized tanks and vehicles. All of us moving steadily toward the Iraqi border. We had been rolling north for hours in the cold February air, the monotony broken only by the occasional crackle of radio chatter and the persistent whine of engines. When the order came to halt for refueling, I was grateful for the chance to stretch, even if it meant standing in the biting desert wind.

I climbed out of my vehicle, my boots hitting the sand with a soft crunch. The cold cut through my uniform—that desert cold that people back home never believed was real. The isolation I had felt since that night on guard duty hung around me like a second skin. Quest was ahead, driver of his Humvee. Boyfriend was somewhere in another vehicle. Neither had spoken to me since Boyfriend arrived. The silence between us was its own kind of desert—vast and unforgiving.

The refueling operation was routine and mechanical. Soldiers moved with practiced efficiency, breath visible in the cold air, fuel cans and hoses appearing as if by muscle memory. I stayed near my vehicle, hugging myself against the chill, watching the organized chaos unfold. That's when I noticed something off.

Quest stood beside his Humvee, and even from a distance, I could read the body language. His shoulders had that particular slump of someone who's just realized they've made a terrible mistake. His hand moved to the back of his neck—that universal gesture of "oh shit." Someone was talking to him, pointing at his fuel tank. I could not hear the words over the wind, but I did not need to. Quest had used regular gasoline instead of diesel.

It was an honest mistake, the kind that could happen to anyone in the exhaustion and routine of convoy operations. But it was also the kind of mistake that could have catastrophic consequences. A diesel engine running on gasoline—I did not know all the technical details, but I knew enough to know it was bad. Really bad.

The word spread quickly through our section. The 2nd ACR convoy kept moving, their vehicles disappearing into the bleak horizon ahead of us. We stayed behind. Quest looked sheepish, embarrassed in a way I had rarely seen him. This was a man who always seemed so confident, so capable. Now he stood there, shoulders hunched against the cold, while mechanics and NCOs gathered around his Humvee like surgeons around an operating table.

The directive was clear and urgent: Do not turn that ignition. Do not start that engine. I watched from my vehicle as they worked to drain the fuel tank, every drop of that wrong fuel having to come out before we could move again. The cold desert wind kept blowing. Time stretched. Quest stood nearby, trying to look useful but mostly just looking miserable, his breath coming out in small clouds.

There was something almost humanizing about it. Here was someone who'd also been caught up in the mess I had created—someone who'd made me promises in the dark that neither of us could keep. Someone who wouldn't look at me now. And now he was the one standing exposed, the one who'd messed up. I did not feel satisfaction at his mistake—I was not that petty, and I still cared about him even though everything between us had become impossible. But I did feel a strange sense of recognition. We were all just trying to get through this, all capable of mistakes that could have consequences we could not fully imagine.

What would have happened if someone hadn't noticed? If Quest had driven off with that gasoline sloshing in a diesel tank? If he'd turned that key? It would have been bad. Engine damage at minimum. A fire at worst. The kind of mistake that could leave you stranded in hostile territory, or worse.

The draining seemed to take forever. Our small group of vehicles sat isolated in the cold desert while the main convoy disappeared over the horizon. I huddled in my vehicle, grateful for the warmth. Quest paced with his arms crossed, ears and nose flushed red, wearing the tight expression of a man who wished he were anywhere else, watching his mistake get fixed by other people's hands. He never once looked in my direction. Not when the mechanics first gathered around his vehicle. Not during the long, cold wait while they drained the tank. Not when they finally finished and poured fresh diesel in the tank. The wall between us remained intact, built from hurt and broken promises and the reality that I had chosen—or failed to choose—in a way that left him feeling like he did not matter.

Finally, they finished. Fresh diesel fuel went into Quest's tank. The mechanics checked and double-checked everything, their hands red from the cold and the work. When they finally gave the okay to start the engine, I saw Quest's shoulders relax just slightly. We rolled out, racing to catch up with the convoy that had left us behind, trying to make up time in the cold February afternoon.

In the desert, mistakes had weight. Every action mattered. Every choice could cascade into consequences you could not predict. Quest had just learned that lesson standing in the cold while his error was corrected by others. I had been learning that same lesson in a different way—that one night on guard duty, one moment of loneliness and need, could change everything. Could turn someone who'd shown you kindness into someone who could not look at you. Could make you the villain in the story people told. Could leave you isolated in a convoy full of people.

As we drove north to catch up, I stared out at the endless sand and thought about fuel tanks and wrong choices and all the ways things could go catastrophically wrong if you did not catch them in time. Some mistakes you could drain out and fix. Others you carried with you, mile after mile, no matter how far you drove.

ॐ

The convoy kept moving through the afternoon and into evening. More miles. More cold. More isolation. True darkness had fallen by the time the halt order came.

My team was spread across two vehicles. Specialist Burns and Lt. Holdeman were in the MSQ1O3 Charlie, along with most of our equipment. I was in the Humvee with Staff Sergeant Strychland. The convoy formed a circular perimeter in the middle of nowhere—no camp and no tents. We would sleep in our vehicles and move again at dawn.

I watched as soldiers around us pulled out sleeping bags, settling onto the sand beside their vehicles or climbing into cargo areas to rest. Burns and Lt. Holdeman stayed with the MSQ1O3. Others found spots on the ground. I grabbed my blanket—the army-issue one I always kept with me—and climbed into the back of the Humvee. The rear area was just big enough to stretch out if you did not mind being cramped. I spread my blanket on the hard floor, trying to make it as comfortable as possible, and settled in with my rifle beside me.

The hard-top roof blocked out the stars. It was dark inside—just the faint glow of moonlight through the gaps. I could hear other soldiers settling in around us. Voices speaking indistinctly in the near distance. The creak of metal as people shifted in vehicles or on the ground. I pulled my blanket up and closed my eyes, exhausted from the day's drive.

Then I heard him climbing in. Staff Sergeant Strychland did not say anything. Did not ask if there was room. Did not acknowledge me at all. He just climbed into the back of the Humvee and positioned himself in the remaining space. My stomach tightened, but I kept my eyes closed, willing myself to fall asleep quickly. *It's fine,* I told myself. *Just a few hours. Dawn will come. We'll move again. Nothing's going to happen.*

2♥

I had been trying to avoid Staff Sergeant Strychland ever since the drama with Boyfriend and Sonny Quest exploded across the unit. Everyone knew the story by now. The girl who slept with another guy in the desert. The girl who caused problems.

Staff Sergeant Strychland had been one of the ones watching. Smirking. Making comments just loud enough for me to hear. "Guess Adams likes variety." "Wonder who's next." "Better watch yourself around her—she'll fuck anybody." I had kept my head down. Did my duties. Carried the water. Carried the tripod. Took every punishment he handed out without complaint. Because what choice did I have? He was my team leader. My superior. The one who controlled every aspect of my life out here.

But I had never been alone with him like this. Not at night. Not in an enclosed space where no one could see. I tried to push the thought away. *Go to sleep,* I told myself. *Just go to sleep.*

I must have dozed off. Because when I opened my eyes, everything had changed. Staff Sergeant Strychland was moving. Not settling. Not shifting to get comfortable. Moving toward me. My heart started pounding. Before I could speak, before I could even process what was happening, he was on top of me.

His weight pinned me down, crushing me against the hard floor of the Humvee. I could not move. Could not breathe. He did not say a word. No explanation. No pretense. No question. Just his hands, already at my belt, working it open with practiced efficiency. "Sergeant—" He ignored me. Unbuckled my belt. Unzipped my pants. "Please—" His hand pushed past my waistband, rough and invasive, sliding inside my panties.

My mind was screaming, but my body had frozen. He did not ask. Did not speak. Did not even look at my face. He just assumed. Assumed I would let this happen. Assumed I had no choice. Assumed that fear or duty or shame would keep me silent and compliant. Like I was his to take. Like I was owed to him.

His fingers were touching me now—touching places no one had the right to touch without permission. And in that moment, I saw something. A vision, clear and terrible, flashing in front of me like a prophecy. I saw my future self. An older woman who had been raped in a war zone by a man who was supposed to protect me. I saw myself broken. Destroyed. A shell carrying shame that would never wash off. I saw myself trying to tell someone—anyone—and watching them not believe me. I saw the rest of my life shaped by this one moment, by this one man who thought he had the right to take whatever he wanted. And in that vision, I saw a girl who never recovered. A girl who let this happen. A girl who became a victim forever.

And I thought: *No. That will NOT be me.*

I grabbed his wrist. The one inside my pants. I grabbed it with everything I had and yanked it away from my body. "No. " My voice cut through the silence like a blade. He froze. "NO! " I said again, louder this time. Firmer. For the first time since he'd climbed on top of me, he looked at my face. And I looked back. I did not cry. Did not beg. Did not apologize. I just held his wrist in my grip and stared him down. *Try it,* I thought. *Try to keep going and see what happens.* I did not know what I would do if he forced it. Did not know if I would scream or fight or claw his eyes out. But I needed him to believe I would. Needed him to see that I was not going to just let this happen. That this girl—this nineteen-year-old Black girl from DC who everyone thought was weak—was not going to be his victim.

For a long, terrible moment, neither of us moved. I could feel him deciding. Weighing his options. Wondering if he could force it anyway. But something in my face must have convinced him. Because he pulled his hand back. Slowly. Like he was doing me a favor. He did not apologize. Did not explain. Did not say a single word. He just rolled off of me, adjusted his uniform, and turned his back.

I stayed frozen against the side of the Humvee, my heart hammering, my hands shaking, my pants still unzipped. I pulled them closed. Buckled my belt with trembling fingers. Tried to breathe. He still did not speak. Did not look at me. Did not acknowledge what had just happened. The silence was suffocating.

I sat there in the dark, clutching my blanket, trying not to cry, trying not to scream, trying not to fall apart. If I fell apart now, he'd win. And I could not let him win.

∂❧

When dawn finally came, the unit got ready to roll out again. Staff Sergeant Strychland climbed out of the Humvee without a word, without looking at me, like nothing had happened.

I gathered my gear and got out as well, feeling uncomfortable as hell after what happened with him in the back of that Humvee. I kept wondering what sparked that. Replaying the previous days' interactions and it was completely ordinary. Nothing transpired between us that would lead to him doing that bullshit. Maybe that's exactly what he thought I was—someone he could use for a quick hump and then go right back to dogging me out with extra duties and silence. The lowest-ranking Black girl on the team. Disposable. Not worth respect before, during, or after. But I let it go in and out of my mind. I really did not have the capacity at this point.

I was gathering my gear when he called out. "Adams." I turned to see him standing with Staff Sergeant Bleach-Bottle Blond, a sergeant I barely knew from another team. My stomach dropped. I walked over, my rifle slung across my shoulder, my blanket clutched in my other hand.

"You're switching teams," Staff Sergeant Strychland said flatly. "Report to Staff Sergeant Bleach-Bottle Blond. Effective immediately." That was it. No explanation. No discussion. Just a transfer.

Staff Sergeant Bleach-Bottle Blond looked at me with cold eyes, like I was a problem she had just been handed. "I better not have any issues with

you," she said. I opened my mouth to respond, but nothing came out. What could I say? She had already made up her mind about me. They all had.

"Get your stuff and load up with my team," Staff Sergeant Bleach-Bottle Blond said, jerking her head toward a different vehicle. "We roll out in ten." I nodded and walked away, feeling both their eyes on my back.

I did not know it then, but that morning transfer changed everything. Not just because of what Staff Sergeant Strychland tried to do in the darkness, but because I said no. Because I refused to become the girl in that terrible vision. Because I chose survival over silence; even knowing it would cost me.

And it did cost me. The retaliation was only beginning.

Chapter 15: Retaliation

Mid February 1991. Remote Saudi Arabian Desert heading for Iraq.

We had just arrived at the new jump location—another stretch of featureless desert, another temporary stop on the way toward Iraq. The team was still unloading gear from the vehicles when she called me over.

The first thing Staff Sergeant Bleach-Bottle Blond did was hand me a duty roster. "Adams," she said, clipboard in hand. "Berm duty tomorrow. 0400 to 0600." I looked down at the roster she had thrust into my hands. My name was at the top. Under the letter A. Adams. "Yes, Sergeant," I said quietly. She walked away without another word.

At first, I did not think much of it. Berm duty was terrible—lying in the cold predawn desert in the prone position, watching for threats that never came—but everyone had to do it. It was just part of the rotation.

But then we jumped to a new location a few days later, and Staff Sergeant Bleach-Bottle Blond posted a new duty roster. My name was at the top again. Adams. And then we jumped again. And again. And every single time—every single location—my name was at the top of the list. While soldiers with last names starting with B, C, D, E, F—soldiers who should've been rotating through duties just like I was—never seemed to get assigned. The roster never made it very far down the alphabet before we would jump again and it would reset. It was always me at the top of the list.

By the third jump, I understood what was happening. We moved locations constantly during this phase of the operation—sometimes staying in one place for less than a week before jumping forward again. Six, seven, eight times we relocated, following the 2nd ACR as they pushed deeper into enemy territory. And every single time, Staff Sergeant Bleach-Bottle Blond started the duty roster over at A.

Not continuing from where the last rotation left off. Not giving anyone else a chance. Just starting fresh. Which meant I pulled duty at every single location.

I knew it was intentional. I knew she had gotten a report about me from Staff Sergeant Strychland—whatever lies he had told her had poisoned her against me before I even showed up. But I did not say anything. Did not ask questions. Did not complain. I just did the work, because what was the point? She had already decided who I was and nothing I said would change her mind.

But it was not just berm duty. It was everything. Water detail? Adams. Shit detail? Adams. KP? Adams. Extra guard shifts? Adams. While Private First Class (PFC) Becky—the other woman on the team—seemed to have all the time in the world to hang out with Staff Sergeant Bleach-Bottle Blond, make phone calls home, and just relax. I never got a break. Never got a day off. Never got to be anything other than the soldier at the top of the list.

❧

The worst part was watching Staff Sergeant Bleach-Bottle Blond and PFC Becky together. They were always laughing. Always talking. Always going off to do things together like they were friends, not a sergeant and a private. Just like when we first arrived in Riyadh, when we were in Tent City at King Khalid Military City (KKMC), I would see them heading to the phone center while I was on shit detail. Now, in the middle of nowhere, I would see them sitting outside the tent, sharing snacks from care packages, while I was pulling back-to-back guard shifts.

And I would think: *That could've been me.* If I had been white. If I had been blonde. If I had been someone Staff Sergeant Bleach-Bottle Blond saw as worthy of friendship instead of suspicion. But I was not. So I got the duties. And they got the privileges.

I kept my head down. Did the work without complaining. Stayed quiet. Tried to make myself invisible. But it did not matter because

Staff Sergeant Bleach-Bottle Blond had already decided who I was. And nothing I did was going to change her mind.

ॐ

But here's what Staff Sergeant Bleach-Bottle Blond did not understand: By sending me away on all those duties, she was actually giving me a gift. When I was on KP detail, I was not around her resting bitch face. I was with the 2nd Armored Cavalry Regiment (ACR) cooks—mostly Black guys who welcomed me with open arms, who called me "Cocoa" because I always asked for everyone's hot chocolate packets, who made sure I got extra helpings when there was real food instead of MREs. They treated me like family. Made me laugh. Made me feel human again.

When I was on burn detail, I was not isolated in my own platoon where everyone knew the gossip about Boyfriend and Sonny Quest. I was with soldiers from other units who did not know my story, did not care about the drama, who would joke around even while we stirred barrels of human waste and burning fuel. We would make the best of it. Find humor in the absurdity. Have actual conversations.

And when I was on water detail or guard duty or any other assignment that took me away from my team, I was not around Staff Sergeant Strychland's hateful scowl or Staff Sergeant Bleach-Bottle Blond's disdain and judgment. I was free.

So yes, my body ached. Yes, I was exhausted. Yes, the work was brutal and unfair and meant to break me down, but in a strange way, those duties saved me. They kept me away from the people who were trying to destroy me. They gave me moments of actual connection with people who saw me as a person, not a problem. They let me breathe. And they kept me occupied enough that I did not have to face Boyfriend's hurt eyes or Sonny Quest's anger, or the whispers that followed me everywhere else.

One day after KP duty, I collapsed on my cot and actually smiled. The cooks had made real sausage gravy and biscuits that morning, and I had eaten until I was stuffed. My hands were rough from work.

My body was tired, but I had spent the day with people who liked me. Who did not judge me. Who did not know or care about my mistakes.

And I thought: *Maybe this is how I survive. Not by fighting the punishment. But by finding the blessing hidden inside it.*

I was alone in my platoon. But I was not alone everywhere. And somehow, impossibly, I kept going.

Chapter 16: Shock and Awe (Battle of 73 Easting)

February 26, 1991. Eastern Iraq, near the Iraqi-Kuwaiti border. The day I thought I was going to die.

We had been moving for hours. Our convoy had been pushing forward, following the 2nd Armored Cavalry Regiment as they advanced toward the Republican Guard. We were attached to them for this operation. My unit, Bravo Company, 511th Military Intelligence Battalion, provided intelligence support as they moved deeper into enemy territory.

It was late afternoon when we stopped. Not a planned stop. Not a rest break. We stopped because there was a battle happening in front of us.

I didn't see it at first. From inside our Humvee and our MSQ103 Charlie —all I could see was the desert stretching out in every direction, flat and featureless. But then I heard it. The sound of artillery fire. Distant at first. Like thunder rumbling on the horizon. But getting closer. And louder.

"What is that?" I asked, my voice tight. Staff Sergeant Bleach-Bottle Blond, sitting in the front of the Humvee, did not answer right away. She was listening to the radio, her face tense. "Contact," she finally said. "Friendly forces engaged with enemy armor." Enemy armor. Tanks. We were rolling into a tank battle.

The convoy slowed, then stopped completely. "Everyone out," Staff Sergeant Blond ordered. "Find cover. Now!" We scrambled out of the vehicle, grabbing our weapons, our gear, trying to figure out what "cover" meant in the middle of a flat desert with no trees, no buildings, no shelter of any kind.

"Dig in!" someone shouted. Dig in. That meant foxholes. I grabbed my entrenching tool—that short-handled shovel we were issued— and started digging. The ground was hard. Not soft sand like I had expected, but packed earth mixed with rocks and gravel. Every strike of the shovel sent a jarring shock up my arms. I dug as fast as I could, but it was not fast enough.

The hole was barely a foot deep when the first explosion lit up the horizon. I froze, staring at the orange fireball rising into the sky, maybe three city blocks away. Maybe closer. "Keep digging!" someone yelled. I dug. My arms burned. My hands cramped around the handle of the tool. Sweat poured down my face even though the sun was starting to set and the temperature was dropping. But I kept digging. Because I knew—God, I knew—that this shallow, pathetic hole might be the only thing between me and death.

By the time the sun went down, my foxhole was maybe two feet deep. Not deep enough. Not even close. But it was all I had. I climbed in, pressing my body as flat as I could against the bottom, and waited.

The battle started in earnest as darkness fell. And it was like nothing I had ever experienced. Nothing I had ever imagined. In the movies, battles are loud and chaotic, and you can see what's happening. But this was not like the movies. This was pitch black. The kind of darkness that swallows everything. No moon. No stars. Just endless, suffocating blackness.

And in that darkness, all I could see were the fireballs. Red and orange streaks tearing through the night as tanks fired at each other. Mortars exploding in bursts of light that lit up the desert for a split second before plunging everything back into darkness. Tracer rounds arcing across the sky like deadly fireworks.

And the sound. The sound was unimaginable. Every explosion was a physical force—a pressure wave that hit you in the chest, that made the ground shake, that made you feel like the world was ending. And it wouldn't stop. KABOOM! Boom! BOOM! Over and over and over.

I lay in that foxhole, my hands pressed over my ears, my rifle beside me, and I prayed. *Please, God. Please don't let me die. Please don't let a shell land on me. Please keep me safe.* I prayed the same thing over and over, like a mantra, like if I said it enough times God would have to listen.

I did not know who was winning. Did not know if the fireballs I was seeing were friendly or enemy tanks exploding. Did not know if we were advancing or retreating or just holding position. It became obvious that death was everywhere around me. Close enough to feel. Close enough to hear. Close enough to smell—that acrid, chemical smell of explosives and burning metal. And I was lying in a hole in the ground, terrified, waiting for it to be over.

At some point during the night, I heard someone moving nearby. Boots crunching on gravel. I gripped my rifle, my heart pounding. "Adams?" It was Boyfriend. Boyfriend? "Yeah," I called back, my voice shaking. "You alright!?" "Yeah." "Stay down," he said. "Don't move unless you have to." "Okay." His footsteps moved away, and I was alone again. But knowing he was out there—checking on people, making sure we were okay—made me feel just a little less terrified.

The battle went on for hours. I don't know how many. Time stopped having meaning. There was only the darkness and the explosions and the fear. And the prayer. *Please, God. Please.* Sometime before dawn, the explosions started to slow. Not stop. Just... slow. The gaps between blasts got longer. The sound of artillery fire grew more distant. And finally, mercifully, it stopped.

The silence that followed was almost as terrifying as the noise had been. Because I did not know if it was over, or if this was just a pause before it started again, I stayed in my foxhole, not moving, barely breathing, until the sky started to lighten. First a dark blue. Then gray. Then the faintest hint of orange on the horizon. Then dawn. I had made it through the night.

❧

When it was light enough to see, I climbed out of my foxhole. My legs were stiff. My hands were numb. My whole body ached from lying in the same cramped position for hours. But I was alive. I looked around. Other soldiers were emerging from their foxholes too, blinking in the early light, looking as shell-shocked as I felt.

And then I saw it. The aftermath. The desert in front of us looked like the end of the world. Burned-out tanks—dozens of them—scattered across the sand like discarded toys. Some were still smoking. Some were still burning, flames licking at twisted metal. Vehicle hulks, blown apart by artillery fire, reduced to blackened skeletons. Craters in the ground where mortar rounds had hit. Debris everywhere—shell casings, twisted metal, things I did not want to identify.

And the smell. Burning rubber. Burning fuel. The acrid smoke from burning tanks. I pulled my scarf over my nose and mouth as we drove past. There were bodies in those tanks. Burning. I knew it even though I could not see them. I tried not to imagine what it had been like for them. Tried not to think about how close we had been to ending up the same way.

"Mount up!" someone shouted. "We're moving out!" We climbed back into our vehicles, and the convoy started rolling forward. Slowly. Carefully. Because we had to drive through the battlefield. Past the burning tanks. Past the wreckage. Past the evidence of just how close we had come to dying.

I sat in the back of the Humvee, staring out at the destruction, and I felt... numb. Not relieved. Not grateful. Just numb. Like my brain could not process what I had just survived. Could not reconcile the fact that I was still here, still breathing, still alive, when so many other people—Iraqi soldiers, probably some of them not much older than me—were dead.

༄

Later, I would learn what that battle was called. The Battle of 73 Easting. One of the largest tank battles in American history. A decisive victory for the 2nd Armored Cavalry Regiment. The battle that broke the back of the Iraqi Republican Guard.

But in that moment, I did not know any of that. I just knew that I had been there. In firing range of mortar rounds, tanks exploding and men dying. That I had survived it in a foxhole I could barely dig. And that I had never be the same.

I remembered those white boys back in Germany—the ones who'd been fist-pumping and shouting "Hooah!" when they heard we were deploying. The ones who'd been so excited about getting some "real combat action." The ones who'd talked about war like it was a video game, like it was something to be won. Well, they got what they came for. I wondered if they were still excited. If they still thought this was glorious. If the reality of burned bodies in tanks and the smell of death and the way your hands shake for hours after matched whatever fantasy they'd been carrying in their heads. I doubted it, but they'd never admit it. They would go home and tell war stories. Brag about being in combat. Wear their 2nd ACR patches with pride. And that was fine. Let them have it. I just wanted to survive.

But let me be clear: I was there too. Attached to the 2nd ACR as they fought. Not safely behind the lines—there were no lines. Just desert and darkness and death close enough to touch. If one of those mortars had landed just a little differently, if the battle had shifted just slightly,

I would be dead.

Chapter 17: Iraqi Surrender

Iraqi Desert. Late February 1991.

We encountered them two days after the Battle of 73 Easting. Our convoy was moving north through Iraq—still following the 2nd ACR, still pushing deeper into enemy territory—when someone in the lead vehicle spotted movement ahead. The convoy slowed. Then stopped.

"What is it?" I asked, craning my neck to see. Staff Sergeant Blond was listening to the radio again, her face unreadable. "Iraqi soldiers," she said finally. "On the highway up ahead." My stomach dropped. "How many?" "A lot." We waited in the vehicles for what felt like an eternity. The radio crackled with terse communications—commanders trying to figure out what was happening, whether this was an ambush, whether we should advance or hold position. And then the word came down. Staff Sergeant Bleach-Bottle Blond was listening to the radio, her face tense. "They're surrendering," she said finally.

I did not believe it at first. Surrendering? After everything we had been through—the months of build-up, the artillery strikes, the ground campaign, the tank battles—they were just... giving up?

And that's when I saw them. There must've been eighty of them. Maybe a hundred. Iraqi soldiers, lined up on both sides of the highway, their weapons laid on the ground in front of them. And they were bowing. Not just kneeling. Bowing all the way down, foreheads to the pavement, in full prostration. Like they were praying. Or begging.

I stared from the vehicle, frozen. This was the enemy. These were the soldiers we had been told were monsters. Sadistic. Brutal. The ones who'd invaded Kuwait, who'd set the oil fields on fire, who Saddam Hussein had sent to fight us. But they did not look like monsters. They looked like boys. Thin. Dirty. Terrified. Some of them could not have been more than sixteen or seventeen years old. And they were surrendering. En masse. Without a fight.

Our convoy slowly rolled past them, and I watched from the vehicle, my heart pounding. Part of me was terrified this was a trap. That they would suddenly reach for hidden weapons, that this was some elaborate ambush and that we were all about to die. But they did not move. They just stayed there, bowed down, waiting.

"They're starving," someone said quietly. I looked over. It was one of the guys from another team, staring out at the surrendering soldiers. "Saddam was not feeding them," he continued. "Barely giving them any supplies. They've been out here for months with nothing."

I looked back at the Iraqis. Now that I was closer, I could see how thin they were. How ragged their uniforms looked. How some of them were barefoot. These weren't elite soldiers. These were conscripts. Kids who'd been forced to fight a war they did not believe in, for a leader who did not care if they lived or died. And now they were giving up, because they knew we would treat them better than their own military.

We did not stop. The convoy kept moving, and we left the surrendering soldiers behind. Other units would process them. Take them as prisoners of war. Feed them. Give them medical care. But we had to keep moving. The war was not over yet.

※

Later that night, at our new jump coordinates, I could not stop thinking about the Iraqi soldiers. About the looks on their faces. About how young they were. About the fact that if things had gone differently—if Saddam had equipped them better, if they'd been given more food and supplies, if they'd actually wanted to fight— they might have killed me. Or I might've killed them, but instead, they'd just... given up. And I did not know how to feel about that.

Part of me was relieved. Every Iraqi soldier who surrendered was one less person trying to kill us. One less threat. One less reason to be afraid. But another part of me felt... sad. Because these were just people. People who had been caught up in something bigger than themselves, just like I had. People who did not want to be here, just like I did not want to be here. People who just wanted to go home.

I replayed the propaganda we had been fed before the war started. About how the Iraqis were ruthless, how they'd fight to the death, how they'd use chemical weapons on us without hesitation. And yeah, Saddam Hussein was a monster. Yeah, his regime had done terrible things. But these soldiers? These kids lined up on the highway, bowing to the ground, begging for mercy? They weren't monsters. They were just scared. Just like me.

That night, I lay on my cot, staring at the canvas ceiling of the tent. I kept seeing their faces. The Iraqi soldiers. How young they looked. How desperate. I kept thinking about how easily it could've been the other way around. How if I had been born somewhere else, under a different flag, with a dictator giving orders—I could've been one of them. Hungry. Abandoned. Surrendering to an enemy I had been told would kill me.

War was not what I thought it would be. It was not about good guys and bad guys. It was just about people trying to survive. And most of us—on both sides—were just kids who'd been sent to kill each other by old men who'd never have to see what we saw.

I closed my eyes and prayed. Not for victory. Not for glory. Just for this to be over, so I could go home.

Chapter 18: Swallowed by the Storm

Iraq-Kuwait theater. Late February 1991.

After the surrender, we stayed on the move. Not advancing anymore—the war was essentially over—but repositioning. Following orders. Jumping from one temporary location to another as the higher-ups figured out what to do with thousands of soldiers, Iraqi prisoners, and the wreckage of a hundred-hour ground war.

We would set up camp, stay a few days, then pack up and move again. It was exhausting in a different way than combat. Not the sharp terror of mortars and tank fire, but the dull grind of uncertainty. Waiting for orders. Doing whatever duties they assigned to keep us busy. KP duty. Guard duty. Equipment maintenance. The usual rotation of tasks that fill the endless hours between war and going home.

᠎᠎

I was on my way to another KP shift when it happened. It happened without warning. One minute, I was walking across camp toward the KP tent, just another evening duty assignment, nothing out of the ordinary. The next minute, the world turned red.

I did not see it coming. Did not feel the wind pick up gradually or notice the sky changing color. One second it was normal— that dusty, hazy desert evening light. The next second, the air was thick with sand. A sandstorm. But not like anything I had seen before. In fact, I had never seen a sandstorm in real life. Just heard about them. This was complete. Total. Blinding.

The sand swirled around me in every direction, so thick I could not see my own hands when I held them in front of my face. The wind howled, pelting my exposed skin with tiny grains that stung like needles. I immediately wrapped my wool scarf tight around my mouth and nose, leaving just a slit for my eyes. My goggles blurred with dust within seconds. The world disappeared into a wall of red and brown.

Panic hit me like a physical blow. I spun around, trying to orient myself, trying to figure out which direction I had been walking. But I could not tell. Everything looked the same. Just swirling red sand in every direction. No landmarks. No tents. No people. Nothing. I was completely and utterly lost.

I took a step forward. Then another. *Where am I? Which way is the KP tent? Which way is my own tent?* The storm stripped away my bearings. The camp was not that big. If I could just pick a direction and walk, I would eventually hit something. A tent. A vehicle. Another soldier. But what if I walked the wrong way? What if I walked out into the open desert, away from camp, and got so turned around I could not find my way back? What if I collapsed out there and no one found me until it was too late?

My heart was racing now, pounding hard against my ribs. I was terrified. About as terrified as I had been during the Battle of 73 Easting. More terrified than I had been in the gas chamber back at basic training, because this was different. This was slow. Disorienting. Suffocating in its completeness. This was dying by inches, not knowing which way was safe, not able to see or think or do anything except stand there and wait for it to be over. And I did not know if it would be over in five minutes or five hours.

I'm going to die out here. The thought hit me with sudden, terrible clarity. *I'm going to die, and no one will know where I am.* Instinctively, I started praying. God had always protected me, and I needed Him now more than ever. I could not let panic take over and make things worse. So, I forced myself to breathe (through my scarf). To calm down. To call on Him. He'd always been there—even in my darkest moments, even when fear threatened to paralyze me.

"Please," I whispered into the fabric covering my face. "Please help me. I don't want to die out here. Please. I need help. I don't know where I am. I can't see. I can't—"

And then—I don't know how to explain this except to say it's true—I felt the Holy Spirit talking to me in my head. Like someone was walking with me, leading me, even though I could not see them. The Holy Spirit spoke, calm and steady, cutting through the chaos around me. *Calm down. Focus on Me. Don't think about the storm. You've walked this path so many times before. You know it by heart. Look with your mind's eye.*

Not an audible voice. And not a vision. Just a sense. A knowing. And suddenly, I could see it—not with my eyes, but in my memory. The route I had walked dozens of times. Every mundane trip to KP duty had etched a mental map in my mind. From the tent area: thirty paces forward, slight angle left. Past the latrines: twenty more steps, bear right. Toward the KP tent...

The Holy Spirit said, *"Walk forward. Retrace your steps."* So I did. I started counting. Retracing the path from memory. One step. Two steps. Ten. I could not see where I was going. I could not see anything except swirling red sand. But my feet knew the way. And with every step, I felt it. That presence. That guidance.

I don't know how long I walked. Maybe five minutes. Maybe ten. It felt like an eternity. But then, suddenly, there was a shape in front of me. Blurry. Indistinct. But a shape. My hand reached out—not searching but knowing. Muscle memory from all those times I had pushed through that door. I could not see the handle. But my hand found it anyway. Exactly where it should be. I grabbed it and pulled.

I stumbled inside and shut the door behind me. And the world went quiet. I stood there for a moment, shaking, trying to process what had just happened.

And then I heard them. "Cocoa!" They called out in unison—all of them, together—like a chorus welcoming me home. I looked up. I was in the KP trailer. The 2nd ACR cooks—my cooks, the ones who'd been feeding me and looking out for me—were all there, and their faces lit up when they saw me.

It melted my heart. In that moment, I felt celebrated. Protected. Seen. Like I had just survived an attack by the enemy and they were rejoicing that I had made it through alive. These men did not know about the drama I had with Strychland or Bleach-Bottle Blond or all the ways I had been broken down. But they knew I mattered. And hearing my name—my Cocoa name, the one they'd given me—shouted in unison like that felt like love.

"You made it!" one of them said, his grin wide. "How the hell did you find us in that storm?" I opened my mouth to answer, but no words came out. Because I did not know how to explain it. Did not know how to tell them that I hadn't found them. God led me here, through a blinding sandstorm, when I could not see, could not think, could not do anything except trust. He'd led me here.

"Something happen out there?" one of the cooks asked, his smile fading as he saw my face. I pulled down my scarf but didn't answer. I didn't have the words. I was shaken. Shaken in a way that had nothing to do with the storm and everything to do with the realization that I had just experienced something I could not explain. Something miraculous.

I sat down on an overturned crate, my hands still trembling, and one of the cooks handed me a bottle of water. "That storm came out of nowhere," he said. "We barely made it inside ourselves." I nodded, sipping the water, trying to calm my racing heart. And I thought: I could've died out there. I should've died out there. But I didn't. Because God had heard me. Had answered me. Had carried me to safety when I could not save myself.

The storm lasted another hour. I stayed in the KP tent the whole time, helping the cooks with whatever they needed, grateful to be somewhere safe and warm and surrounded by people who cared about me. When the storm finally passed and the air cleared and I could see the camp again, I walked back to my tent. Quickly. Still shaken.

❧

That night, lying on my cot, I replayed the whole thing in my head. The terror. The complete disorientation. The moment I had cried out to God in desperation. And the way He'd answered immediately. Not with a booming voice or a beam of light or anything dramatic. Just with a quiet knowing. A calm presence. A sense of direction when I had none. *You know the way. Walk forward.* And I did. He led me through that storm when I could not see a foot in front of my face. He brought me to safety when I was completely lost. He saved me. Again.

I lay there in the darkness, and something broke open inside me. Shame. Deep, crushing shame. Because I finally understood. God had been with me this whole time. Through basic training. Through Germany. Through the deployment.

Through the battle. Through the assault. Through everything. He'd been there. Watching over me. Protecting me. Trying to get my attention. And I had been running. Not seeking Him. Not being obedient. Not turning to Him the way I should have turned to Him.

It took a sandstorm in the middle of a war zone—blind, lost, terrified—for me to finally cry out His name. It took being stripped of everything—my sight, my sense of direction, my control—for me to finally surrender.

I wept that night. Tears of contrition. Of repentance. Of shame. But also tears of gratitude. Because He answered anyway. Even though I had been distant. Even though I had been disobedient. Even though I hadn't been seeking Him the way I should have sought Him. He still answered when I called. He still led me to safety. He still loved me.

I'm sorry, I prayed into the darkness. *I'm so sorry. You've been trying to reach me for so long, and I was not listening. I was too proud. Too stubborn. Too caught up in everything else. But I'm listening now. I see You now. Please forgive me.*

And in that tent, in the desolate Iraqi desert, surrounded by people who wished I did not exist, I finally turned my heart fully toward God. Not out of habit, or tradition, or because it's what I was supposed to do. But out of complete humility. Out of recognition that I was nothing without Him. That every breath I had taken, every moment I had survived, every time I should have died, but didn't—that was Him. God had saved me. Again, and again, and again. And I had been too blind to see it. Until now.

For the first time in a long time, I felt something other than fear. I felt faith. Not the kind of faith I had before—untested, theoretical, something I believed in but had never had to rely on to survive. This was faith that had been tested in the fire.

The kind that comes from experiencing something you can't explain. The kind that comes from crying out in desperation and hearing an answer. The kind that comes from being lost in a storm and being carried home through darkness. The kind that changes you. God had led me through that sandstorm. And somehow, I knew—He would lead me through everything else too.

Chapter 19: Ceasefire and Certainty

Kuwait Desert. February 28, 1991.

The ceasefire was announced, and just like that, the war was over. A hundred hours of ground combat. Thousands of Iraqi soldiers surrendering. Oil fields burning black across the Kuwaiti sky. And now—silence. We won. But winning did not mean going home. It meant waiting.

We were still in Kuwait—some remote staging area in the middle of nowhere, surrounded by the wreckage of war and those burning oil fields blackening the sky. Some soldiers handled it better than others. They played cards. Wrote letters home. Worked out. Found ways to fill the endless hours between combat and going home.

I read, as books had always been my escape. During those three weeks in Kuwait, books became my lifeline. I devoured anything I could get my hands on—mostly V.C. Andrews or Danielle Steel romance novels that had been donated and passed around. They were both prolific authors with endless sagas that kept me engrossed for hours. The J.K. Rowlings of the nineties, before Harry Potter existed. Page-turners. Pure escapism. Stories about love and drama and lives that had nothing to do with war. For a few hours at a time, I could disappear into those pages and forget I was in Kuwait. Forget the sand. Forget the smoke. Forget the fear. Forget everything.

I did not think about Boyfriend or Quest much. That whole mess—the guard duty night, Boyfriend showing up the next day, the chaos that followed—felt like it had happened to someone else. I had hurt them both. Deeply. And there was no fixing it. So, I kept my distance. Stayed out of their way. Let the silence stretch between us. It was easier that way.

ॐ

That's when I realized. Not all at once. Not with some dramatic revelation. Just slowly, over the course of a week, as my body kept telling me something I did not want to believe. I should have gotten my period by now. I had been tracking it in my head—counting days, doing the math, trying to convince myself it was just the stress, just the deployment, just my body reacting to everything we had been through. But deep down, I knew. I had an inkling for weeks, actually. Maybe since that night in the Humvee with Quest back in late January. Maybe even before that.

During all those weeks of active combat—through Iraq, through Kuwait, through the Battle of 73 Easting when I lay in that foxhole certain I would die, through the sandstorm that blinded me completely, through those burning oil fields that turned day into night—I had been careful. I had wrapped my scarf around my face whenever I had to go outside. Held my breath when the wind shifted and brought the black smoke toward us. Stayed in our staging areas as much as possible, eating MREs in the vehicles or tents, not venturing out unless I had to. I did not know for certain I was pregnant then, but some part of me suspected. And I had protected myself just in case.

Now, in limbo with all the time in the world to think about everything that I had been through in this warzone, while mindlessly staring at endless desert at this remote Kuwait outpost, I could not deny it anymore.

The cramps came first. That familiar ache low in my belly that usually meant my period was starting. I would feel it and think, Okay, here it comes. I would check my underwear. Nothing. Wait a few hours. Check again. Still nothing. This went on for days. A whole week of cramping, of feeling like I was about to start bleeding any second, of checking and checking and finding nothing. Not a spot. Not a trace. Just cramps. And waiting. And Aunt Flo never showed up.

I was lying on my cot one morning when it hit me with absolute certainty. I'm pregnant. Not maybe pregnant. Not probably pregnant. Pregnant. I lay there, staring at the canvas ceiling, my hands resting on my belly while I tried to do mental conception computations. What date was it when we had sex? What day is today? How far along am I? What happens now? We had talked about having a baby like it was a fairy tale. Now it was real, and I had no roadmap for what I was going to do.

I did not tell anyone. Not Dawn. Not the medics. Only God and I knew. I just... kept going. Did my duties. Ate my MREs. Prayed that my uniform would keep fitting, that my body wouldn't give me away, that somehow I would make it through this deployment and get home before anyone figured it out. I was tall and lanky with no curves to speak of. Maybe that would work in my favor. Maybe I could hide it long enough.

Please, God, I prayed. *Please let me hide it long enough.*

✌

But I did think about my mom. I remembered the phone call I had with her after Christmas, right before the new year, before that fateful night on guard duty with Sonny, before Strychland tried to have his way with me, before we had rolled into Iraq with the 2nd ACR, before the Battle of 73 Easting, before everything went to hell.

That was the time when we had access to a sat phone at one of the camps—one of those rare moments when they let us call home for a few precious minutes. I had waited in line for over an hour just to hear her voice. And when I finally got through, that's when she told me she was pregnant. Seventeen years after having my younger brother, Glenn, she was going to have another baby. I was going to have a little brother or sister. I was happy for her. Shocked, but happy.

And then she said something that floored me: "Get pregnant and come home, Darnnell. We can raise our babies together." I did not know what to say. She was telling me to get pregnant on purpose. To get myself sent home. Because that is what happens—if you got pregnant in a war zone, they had to evacuate you. Medical necessity. They would send you back to safety, back to the States, away from combat. It was a way out. The only way out she could think of for her daughter who was stuck in the middle of a war.

The irony was not lost on me. Because getting pregnant in a combat zone could also get you court-martialed. Having sexual relations while deployed was against regulations. Fraternization. Misconduct. A violation of General Order Number One.

What I knew—deep in my gut, in that place where Black women learn to read danger—was that the rules wouldn't be applied equally. Not for me. For someone else—someone white, someone who was valued by leadership, someone whose pregnancy would be seen as an unfortunate situation rather than proof of bad character—they might get a quiet medical evacuation. Sympathy, a plane ticket home and well-wishes.

But for me? The only Black woman in the unit. The girl they already whispered about and vilified. The one Staff Sergeant Bleach-Bottle Blond had been targeting since day one. The one Staff Sergeant Strychland seemed determined to break. For me, pregnancy wouldn't be a medical issue; it would be ammunition.

I could already see how it would play out. Staff Sergeant Bleach-Bottle Blond would write me up. Document every regulation I had violated. She would make sure it was by the book so there would be no way to fight it. Staff Sergeant Strychland would push for the harshest punishment possible. Maybe a field-level Article 15—formal discipline without a trial. But here's the truth: I was already being punished. Strychland had been issuing his own off-the-books Article 15 since the day I arrived. Same penalties. No paperwork. No defense. No record that could be challenged. Maybe worse. In wartime, commanders have broad authority under the Uniform Code of Military Justice to maintain discipline and order. Punishing someone for misconduct or insubordination would be easy. Especially when they are already looking for a reason.

They would call it fraternization. Sexual misconduct. Conduct unbecoming. They would use whatever charge stuck. And I would have no defense, because technically, I had violated the regulations. It did not matter that Sonny Quest and I had been two lonely soldiers trying to find comfort in an impossible situation. It did not matter that I was nineteen and scared and desperate to believe someone cared about me. It did not matter that I had already been targeted and isolated and punished for things that weren't violations at all. The pregnancy would be proof—concrete, undeniable proof—that I was exactly what they'd always said I was. A problem.

I was not being paranoid. I was not being immature or overly pessimistic. I was reading the room. I had watched how they treated me versus how they treated everyone else. I had seen the double standards, the selective enforcement of rules, the way regulations magically appeared whenever they needed to punish me but disappeared when it came to soldiers they liked. I had learned to trust my instincts. And every instinct I had screamed: *They're looking for a reason. Don't give them one. Watch your back.*

So, revealing my pregnancy wouldn't save me. It would give them exactly what they'd been waiting for. A legitimate reason to destroy me—on paper, official and impossible to argue. At minimum, an Article 15 that would follow me forever. But knowing them? They'd push for a court-martial. And a court-martial could mean a bad conduct discharge, maybe dishonorable. The end of any chance at the future I had joined the Army to build. My mom did not know all the damage that could be done to me. She thought pregnancy meant evacuation. Safety. Coming home. She just wanted me out of the desert before something happened to me. It was the most desperate, loving thing she could've said. She did not know my chain of command would see it as an opportunity to further denigrate me.

So I kept my mouth shut. Hid my pregnancy as long as I possibly could. Did what I was told without arguing, without defending myself, without giving them any excuse to look closer. Because I knew—the way Black women know things, the way survival teaches you to know things—that I was one mistake away from a setup. One revelation away from them pulling out every regulation and violation they could find to bury me. And my baby—my innocent, perfect baby who hadn't done anything wrong—would be caught in the crossfire.

It was one more secret I had to carry. One more impossible choice. Reveal the pregnancy and face court-martial or hide it and risk everything if they found out anyway. Damned if I did. Damned if I didn't. But at least if I stayed quiet, I had a chance. A chance to make it home. A chance to protect my baby. A chance to survive this deployment with my future intact.

So I chose silence. And I prayed—harder than I had ever prayed—that God would help me hide the pregnancy long enough.

And sitting there somewhere in Kuwait in early March, I realized the profound irony. I actually did what she said and got pregnant. It made me laugh to myself. Her desperate plan had already happened—but not the way she had envisioned on the phone. I was pregnant in a war zone, alone, surrounded by people who targeted me. And revealing it wouldn't get me sent home safely—it would destroy me.

What I did not know during that phone call in late December was that everything would change in ways I could not imagine. That I would get pregnant just weeks after that conversation. That we had both been carrying our separate storms—her in DC and me in Desert Storm—neither of us knowing what the other was going through. I wouldn't learn her full story until much later, when the mail finally caught up with me.

But what I knew about me—what I knew about women like me, women who'd been raised to survive anything—was that I did not need Quest or Boyfriend to validate my baby or my choices. My baby was mine and we would be fine. God had entrusted this baby to me, and with His help, whatever happened next, I would figure it out. I had survived poverty. Basic training. Targeting and isolation. Combat. Assault. I could survive this too.

Chapter 20: Shit Detail

Saudi Arabia. March 1991.

After three long, agonizing weeks in Kuwait, the orders finally came. We loaded our gear and headed back to Saudi Arabia. Back to waiting. Back to wondering when—or if—we would ever go home.

The 2nd ACR had returned to Germany, and I felt their absence like a weight. Those men who'd treated me like family, who'd let me work in their kitchens and showed me kindness, were gone. Back to their families. Back to normal life. And I was still here. But at least we were moving. We weren't sitting in Kuwait staring at oil fires anymore. We arrived at our new staging area in mid-March—King Khalid Military City, someone said. It did not look like much of a city to me. Just tents and sand and time stretching endlessly ahead. By then, I had known for about two weeks that I was pregnant. My period hadn't come in late February, and the fear, the certainty, the shock—all of it—had set in back in Kuwait. Now, five or six weeks along, I just had to figure out how to survive new duties while hiding my pregnancy.

৵

The heat was climbing fast. When we first arrived in December, the desert had been bitterly cold. Now, three months later, it was pushing ninety, sometimes over a hundred degrees. And I was still doing burn detail—what everyone called shit detail.

Of all the jobs I had been given, this was the worst. Worse than hauling water. Worse than carrying around the fifty-caliber tripod. Worse than lying on the cold ground during berm duty, freezing in the predawn dark. Shit detail was exactly what it sounded like— burning human waste in the desert. And now I had to do it while pregnant.

The latrines were crude plywood boxes with three toilet seats in a row—no doors, no dividers, no privacy. The waste dropped straight down into fifty-gallon drums positioned underneath, accessible from a door at the back. When it was time for shit detail, you opened that door, crouched down, and pulled the full barrel out—carefully. If you tilted it wrong, the contents could slosh. And the last thing you wanted was to contaminate yourself.

For any woman, pulling a fifty-gallon drum of human waste was nearly impossible alone. That's why it was usually a two-person job. And the company was what made it bearable. I got paired with all kinds of soldiers—guys who cracked jokes to pass the time, women who swore this had to be payback for something, even a few soldiers who only knew me as Cocoa. And we laughed about how we had come all this way to fight a war only to stir barrels of burning feces instead. We made the best of it. Talked about home, music, and foods we missed. And for those few hours, I was not the girl with the scandal. I was not the problem soldier Staff Sergeant Bleach-Bottle Blond was trying to break. I was just another soldier doing a filthy job with people who understood.

Once the barrels were dragged fifty yards from camp, we poured diesel and mogas over the top—enough to burn evenly, but not so much that it flared out of control. Then we lit it. The initial flare-up sent a wave of heat straight at you. It was already in the high 90s most days, but standing next to those burning barrels pushed it well over 100, and you felt every degree. Once the flames settled, we stirred with long metal poles, keeping it burning evenly until everything was reduced to ash. It took hours—sometimes three or more.

We wore no protective gear—no masks, no gloves, no suits—just our brown tees and BDU pants. "Stay upwind" was the only instruction, and the wind was unpredictable. One minute it was steady, the next it shifted, and thick, black smoke rolled straight into your face. You coughed, gagged, covered your mouth with your sleeve, and moved constantly so you could avoid inhaling the toxic smoke.

I was hyper-aware of my developing baby's fragile health, and I was extra careful to remain upwind or wrap my spare tee shirt around my face when I could not avoid the smoke invading my airspace.

My partner that day—a guy from another unit whose name I never learned—noticed. "You okay?" he asked, watching me dance around the barrel. "Yeah," I said. "Just don't want to breathe that shit if I don't have to." He laughed. "Smart. I should probably do the same." But he did not know why I was really being careful. Did not know I was protecting more than myself.

Despite the heat, the smell, the toxic smoke—I did not hate shit detail the way I should have. Being assigned every grunt duty imaginable meant I was nowhere near my haters. When I was on shit detail, I was not near PFC Becky's smug smiles or Staff Sergeant Strychland's menacing toxicity or Staff Sergeant Blond's disgusting favoritism. Out there, I worked beside other soldiers who did not know me and who I did not know. Completely drama- and stress-free. And for a few hours, I could just be chill.

Decades later, I would learn about Gulf War Illness and the thousands of veterans who came home sick with symptoms no one could explain. Chronic fatigue, joint pain, respiratory and neurological problems. I knew about the burn pits firsthand—I had stood over them, stirred them, tried not to inhale those noxious toxic fumes. What I did not know back then was what was in them. Not just human waste, but trash, plastics, electronics, and chemicals. Toxins now linked to cancers, lung disease, and birth defects. No matter how much we tried to protect ourselves, many of us still got sick.

And I would think back to those days in Saudi Arabia, standing over barrels of burning waste, breathing through my shirt to keep from inhaling poison. And I would realize I was one of the lucky ones because my daughter was born healthy. Despite the burn pits, the PB pills, the oil fires, and the stress—she was fine. And that wasn't luck. That was God. And maybe it was also those soldiers who looked out for me without even knowing why—the ones who took over stirring when I got too close, the ones who made me laugh when everything else tried to break me.

Back then, I did not know anything about the Gulf War Illness on the horizon. I just knew that shit detail—as awful as it was—was just another terrible but necessary duty that had to be done. In a strange way, it also gave me some reprieve from isolation. Because stirring those barrels with strangers who provided comic relief for a few hours at a time was therapeutic. Our shared misery during those few hours was helping me survive. One shit detail at a time.

Chapter 21: A Stranger's Mercy (First Ramadan)

Saudi Arabia. Mid-March 1991.

The month of Ramadan had started, and the mood in camp shifted. Not among the Americans—most of us barely noticed, until we received the advisories we had been given during formation.

"Ramadan is a holy month for Muslims," the briefing officer had said. "Our Saudi counterparts will be fasting from sunrise to sunset. Be respectful. Don't eat or drink in front of them. Don't offer them food. Understand that they may be hungry and thirsty, but this is important to them." I had nodded along with everyone else, filed it away as just another piece of information to remember, but I did not really understand what it meant. Not until I saw it.

The Saudi soldiers stationed with us were different during Ramadan. Quieter. More focused. They'd go about their duties during the day—working, patrolling, doing whatever tasks they'd been assigned—but you could see the strain in their faces. The hunger. The thirst. And yet they did not complain. They just kept going, with a kind of quiet dignity that I found both admirable and confusing.

I did not understand why they were doing it. Did not understand what the point was of making yourself suffer like that, but I respected it. And I made sure to take my MREs somewhere out of sight, somewhere they wouldn't have to watch me eat. It seemed like the least I could do.

I had been in South West Asia for three months by this point, and I had barely interacted with any Saudi people. The soldiers we worked with kept to themselves, mostly. They were polite, but distant, and I never really had a reason to talk to them beyond basic greetings. So, when one of them approached me after my guard duty shift one afternoon, I was surprised.

"Private," he said, with his heavily accented English, but clear. "You are Adams?" "Yes, sir," I said automatically, even though I did not think he was an officer. He smiled. "Not sir. Just soldier, like you." I relaxed a little. "Okay."

"Tonight," he said, "my unit will have iftar. The meal to break the fast. We would be honored if you would join us." I blinked. "Me?" "Yes," he said. "You are alone much of the time, I think. It would be good for you to share a meal with friends."

Friends. I almost laughed. I did not have friends out here. But I did not say that.

"I don't want to intrude," I said carefully. "Isn't it a...a religious thing?" "It is," he said. "But hospitality is also important. Very important. It would honor us if you came."

I hesitated. Part of me wanted to say no. Wanted to go back to my tent and read my book and be alone like I always was. But another part of me—the part that had been starving for kindness, for inclusion, for any sign that someone saw me as more than a problem—wanted to say yes.

"Okay," I said. "Thank you. I would like that." His smile widened. "Good. Come to the guard post at sunset. We will eat on the roof." The roof? I nodded anyway. "I'll be there."

When my shift ended, I was nervous. I did not know what to expect. Did not know if I would do something wrong, say something offensive, accidentally disrespect their religion. I had never been around Muslims before—at least not that I knew of. And here I was, a nineteen-year-old Black girl from DC, about to have dinner with Saudi soldiers during their holy month.

What am I doing? But I showed up anyway. Because the alternative was going back to my tent. Alone. Again.

Any other time, I would have chosen that without hesitation. I liked being alone. Preferred it, actually. Books and solitude over people and drama—that was my default. But these Saudi soldiers weren't offering me shallow company or fake friendship. They were offering me dignity. Respect. A place at their table during their holy month.

It reminded me of the 2nd ACR cooks—those men who'd treated me like family, who'd let me work alongside them, who'd shown me kindness without expecting anything in return. There had been moments of grace all along this deployment. People who saw me as human when my own unit treated me like a problem. The cooks. These Saudi soldiers. Strangers who showed me more decency than the people who were supposed to be my brothers and sisters in arms.

That kindness meant everything. It kept me going when everything else tried to break me. So when this soldier invited me to iftar, I said yes. Not because I was desperate for company. But because I recognized what he was offering: the same honor and humanity the 2nd ACR cooks had shown me. That was rare enough to be worth saying yes.

The guard post was a two-story structure—one of the few permanent buildings in the area. I had been on duty there before, but I had never been to the roof. When I climbed the stairs and stepped outside, I stopped.

The roof was transformed. There were carpets spread out across the concrete. Large platters of food arranged in a circle. Cushions for sitting. Lanterns providing soft, warm light as the sun began to set.

And the food. Oh, the food. I had never seen anything like it. Whole roasted chickens. Rice pilaf studded with nuts and raisins. Flatbreads still warm from the oven. Dates and figs arranged in neat rows. Hummus and baba ghanoush in ceramic bowls. Skewered lamb. Grilled vegetables. Fresh salads. Yogurt sauce. Honey pastries dusted with powdered sugar. It looked like a feast fit for royalty. And it smelled incredible.

The Saudi soldiers were already there, sitting cross-legged around the food, waiting. They looked up when I arrived, and several of them smiled. "Welcome," the soldier who'd invited me said, gesturing for me to sit. "Please, join us."

I sat down awkwardly, trying to figure out where to put my legs, how to sit in a way that wouldn't be disrespectful. One of the soldiers noticed and said kindly, "However you are comfortable. There is no wrong way." I relaxed a little and settled into a cross-legged position, tucking my uniform around my boots.

They did not eat right away. They waited until the call to prayer sounded—a distant, haunting melody that echoed across the desert. And then, as one, they reached for the dates. "We break the fast with dates," the soldier next to me explained quietly. "As the Prophet did." I watched as they each ate a date, slowly, reverently. And then they began to eat.

They passed the platters around, encouraging me to take as much as I wanted. "Eat, eat," they kept saying, piling food onto the plate I had been given.

I tried everything. The chicken was tender and perfectly spiced. The rice was fluffy and fragrant. The lamb melted in my mouth. The bread was soft and warm and perfect for scooping up the hummus. And the dates—sweet and rich— were unlike anything I had ever tasted. It was the best meal I had eaten in months. Maybe the best meal I had ever had.

As we ate, they talked. Not to me, really—I did not speak Arabic, and most of them did not speak much English—but to each other. And they laughed. Really laughed, in a way I hadn't heard in a long time. It was not the cynical, bitter laughter of soldiers making dark jokes to cope with the war. It was genuine. Joyful. Free.

And sitting there among them, listening to a language I did not understand, eating food I had never tasted before, I felt something I hadn't felt in months. Peace.

After a while, the soldier who'd invited me turned to me and asked, "You are well?" "Yes," I said. "This is amazing. Thank you for inviting me." He nodded. "It is our duty to share. Especially during Ramadan. To feed those who are hungry, to welcome those who are alone—this is important."

I felt tears prick my eyes, and I blinked them back quickly. "I am alone," I admitted quietly. "So... thank you. Really." He studied my face for a moment, then said, "You are not alone now." And something about the way he said it—so matter-of-fact, so kind—broke something open in me.

They did not ask me any questions. Did not ask why I was alone, or why I looked so sad, or what had happened to make me this way. They just fed me. Made sure my plate was never empty. Made sure I felt welcome. And when the meal was over and it was time to go, they thanked me for coming. Like I had done them a favor. Like my presence had mattered.

I walked back to my tent in the dark, my stomach full, my heart lighter than it had been in weeks. And I could not stop thinking about what the soldier had said. *It is our duty to share. To feed those who are hungry. To welcome those who are alone.* That's what Ramadan was about. Not just fasting. Not just sacrifice. But generosity. Hospitality. Taking care of each other.

And they had taken care of me. A stranger. An outsider. A nineteen-year-old girl who had nothing to offer them in return. They'd seen me, and they'd fed me, and they made me feel human again.

୨୬

Years later, I would learn more about Ramadan. About how fasting is meant to cultivate empathy for those who are less fortunate. About how charity and kindness are considered acts of worship during that month. About how sharing iftar with someone—especially someone who is struggling—is seen as one of the highest forms of devotion.

And I would realize: I was the one who was less fortunate. I was the one who was struggling. And they had seen that, without me having to say a word. And they'd helped me.

That night, lying on my cot, I prayed. Not the desperate, terrified prayers I had been praying for months. But a prayer of gratitude. *Thank You, God. Thank You for those men. Thank You for their kindness. Thank You for showing me that there are still good people in the world. That I'm not completely alone. That even here, in the middle of this war, there is still grace.*

And for the first time in a long time, I fell asleep feeling something other than isolation. I felt hope.

Chapter 22: Beach Blanket Bingo and Dysentery

KKMC, Saudi Arabia. April 1991.

Things got strange. The war was over. We had won. But we weren't going home. Not yet. So we waited. And while we waited, the Army tried to keep us occupied.

It started to feel less like a military deployment and more like... summer camp. A really hot, sandy, weird summer camp where everyone wore uniforms and carried weapons. But still a summer camp.

Someone set up a volleyball net in the middle of the compound. Soldiers started playing hacky sack between shifts. There were card games. Weightlifting competitions. People tanning in their PT gear like they were at the beach. Beach Blanket Bingo in the desert. That's what it felt like. I watched the volleyball but didn't play—I had never been good at it, and I was not about to embarrass myself trying. So I cheered when someone made a good play, ate my hamburgers, and enjoyed the rare moment of normalcy.

And then one day, they set up a blister bag pool. An actual pool. Just a huge rubber bladder filled with water, but it was big enough to swim in, and in the April heat—temperatures climbing past 100 degrees now—it felt like heaven. Soldiers were jumping in fully clothed, splashing around, laughing like kids.

I went in too. The water was warm—really warm, but it felt soooo good. It felt amazing just to be submerged, to cool off, to do something normal. For a little while, I forgot where I was and forgot what I was carrying. Forgot everything except the feeling of water on my skin.

They even grilled hamburgers. Real hamburgers. Actual beef, even though we were in Saudi Arabia where cows were sacred. I don't know how they managed it, but they did. And those burgers—

charred and juicy and served on soft buns—were the best Western food I'd eaten since leaving Germany. We ate under big red-and-white checkered tents that had been set up for meals. It almost felt festive. Like a cookout. A celebration. Except we were still in the desert, still waiting for orders, still stuck. But even in the middle of all this, I was still pregnant. Still hiding it. Still dealing with symptoms I could not explain.

The worst part? The constant need to pee. Pregnancy does that—makes you have to go all the time, urgently, without warning. And the latrines were so far away. At least a five-minute walk from my tent, maybe longer. During the day, it was annoying. At night, it was unbearable.

I would wake up in the middle of the night with my bladder screaming at me, and I would lie there on my cot trying to calculate whether I could make it to the latrine or if I should just try to hold it until morning. But I could not hold it. So I would get up, pull on my boots, and start the long walk across the dark compound. Except one night, I realized: I'm not going to make it.

So I did not even try. I grabbed my poncho, stepped outside my tent on the side near my cot, and draped it over my body. It was long—hung down like a dress—and covered me completely. I squatted right there, hidden beneath the poncho. In the sand. Next to my tent. And I peed. It was that, or wet myself.

After that, I did it regularly. Middle of the night, everyone asleep, nobody around. Just me, my poncho covering me like a long dress, squatting in the sand. Quick. Quiet. Done. Nobody ever knew, and thankfully, compared to what was happening to the men, my midnight pee breaks were nothing. Because in early April, a dysentery outbreak hit. And it hit hard.

One by one, the men started getting sick. Explosive diarrhea. Stomach cramps. Fever. Dehydration. Dysentery. And it spread like wildfire through the male soldiers.

The strange thing? None of the women got it. Not one. I don't know why—maybe we were more careful about hygiene, maybe we drank different water, maybe we just got lucky—but every single case was a man. And there were a lot of cases.

You could tell who was sick because they'd be running. Sprinting across the compound, clutching a roll of toilet paper, and holding their stomach with one hand, the other hand holding their buttocks, hoping desperately not to shit themselves before they made it to the latrine. And sometimes, they didn't make it.

I saw grown men—soldiers, warriors, guys who'd been in combat—reduced to running across the sand with panic in their eyes because their bowels were about to betray them in the most humiliating way possible. It would've been funny if it was not so awful. Actually, it was still kind of funny.

And those guys who'd been fist-pumping and yelling "Hooah!" back when we got deployment orders. The ones who were so excited to see "real combat action"—yeah, they were the ones running across the compound with toilet paper in hand, trying not to shit themselves. That's some real combat action.

Boyfriend got it. I saw him one afternoon, running full speed toward the latrines, with a roll of toilet paper in one hand, and a face looking anguished with desperation. He didn't make it. He told me later—and word got around, because of course it did—that he'd had to go on the other side of the berm, almost in full view of everyone, because there was no way he was making it to the latrine. "I was not gonna make it," he said. "I had to go right there or I was gonna shit my pants."

I did feel bad for him. He was still a good guy. Still decent to me, even after everything. The relationship was over, but he hadn't turned mean about it. He was handling it with more grace than I probably deserved. Quest, on the other hand, was sullen and scorned. Barely spoke to me anymore. To this day, he hasn't. But Boyfriend? He was still a gentleman. Still kind. And watching him have to go through that—the humiliation of it—I felt for him. Even if we weren't together anymore.

The dysentery was not just hitting them from one end, either. It was both ends, at the same time. Vomiting and diarrhea, simultaneously, like their bodies were staging a full revolt. Some guys had to make split-second decisions about which emergency to prioritize. There were no good options. The outbreak lasted for weeks. The med tent was overwhelmed. Soldiers were getting IV fluids for dehydration. The latrines were even more disgusting than usual, if that was possible. And the rest of us—the ones who weren't sick—just tried to stay out of the way and not catch it.

Meanwhile, I was still dealing with my own bathroom issues. Still peeing constantly. Still making those long walks to the latrine during the day. Still sneaking out at night with my poncho when I could not wait. But at least I was not shitting myself. Small victories.

We did get one upgrade during this time: hot showers. Not great showers—still just those blister bag contraptions where the sun heated the water—but they were hot, and they were regular, and after months of baby wipes and cold rinses, it felt like luxury. I would stand under that warm water for as long as I could, letting it wash away the sand and sweat and stress. It did not fix anything. But it helped. The latrines were still terrible. Still a long walk. Still disgusting. But at least I could shower after.

⁊❧

Life in Saudi during those weeks was surreal. Beach parties and dysentery. Volleyball and desperate sprints to the latrine. Grilled hamburgers and IV fluids for dehydration. Normalcy and chaos, all mixed together.

I was still getting assigned the grunt work. Shit detail. Guard duty. While Staff Sergeant Bleach-Bottle Blond and PFC Becky went on their girls' trip to the Saudi markets, souvenir shopping and sightseeing, I was hauling water or stirring burn pits or standing watch. Nothing had changed on that front. But I had the pool, the hamburgers, those moments of almost-fun, when I could pretend everything was okay, and I had my secret. Still growing. Still hidden. Still mine.

It wouldn't be long now. We would be leaving soon—going to the final staging area, then hopefully home. I just had to hold on a little longer. Just a little longer.

Chapter 23: Separate Storms, Same Prayer

Qaisumah Airbase, Saudi Arabia. Late April 1991.

The mail finally caught up with me. Not a trickle—a flood.

Canvas bags dumped on a table in the mail tent, soldiers crowding around, shouting names, tossing envelopes across the room. I pushed through the noise and grabbed the stack handed to me—six, maybe seven letters. More mail than I had received during the entire deployment.

I carried them back to my cot, sat down, and stared at them for a moment. Letters from home. From a world that felt distant and unreal.

I sorted through the envelopes—my mother, my grandmother, a couple of cousins—and then I saw one that stopped me.

Monisolo.

My childhood friend from Backus Junior High. The girl I had laughed and dreamed with before the Army changed everything. Her return address said Frostburg University, Maryland. She was living the life I once imagined for myself.

Her letter was written in purple ink, two sheets of notebook paper, and a photo fell out when I opened it—Moni smiling in her dorm room, safe and content.

She wrote that she did not know if her letter would reach me, that she prayed I was safe. She told me she thought about me often, that the girls back home still remembered me. She said I was not forgotten.

I read her words twice. They did not sound like much on paper, but to me, they were life. I had been surrounded by hostility for months during this time, and here was proof that somewhere in the world, someone still cared whether I lived or died.

Moni at Frostburg

Then I opened my mother's letter.

She told me she had given birth on January second. When she was only five months pregnant. The baby was tiny and fragile, fighting for every breath, but alive. My mother had named her Lucky.

She had written me about a month after the birth, but the terror, anguish, and heartbreak of almost losing her beloved daughter was still fresh on the page. And there I was, only three days into her very tragic crisis on January fifth, writing her a "wish list" letter asking for care packages. Hair moisturizer. A book-clip flashlight. Baking soda for my teeth. I had complained about the cold and the duties and asked her to send articles from the Washington Post.

While she was in a hospital holding a premature baby, praying her daughter would survive, I was asking for snacks and toiletries.

Neither of us knew what the other was going through.

My mother had almost lost her daughter. She had spent weeks not knowing if Lucky would survive, watching her tiny body fight for every breath in a hospital incubator. The sleepless nights. The prayers. The terror of losing a child she had waited seventeen years to have.

And I hadn't been there. Hadn't known. Hadn't been able to offer her a single word of comfort.

Then I reached the part that changed everything. She repeated what she had said on the phone before: "Get pregnant and come home. We can raise our babies together."

I had been terrified to tell her. Terrified she would see me as a failure—another inner-city statistic caught up in the cycle of poverty. Another young Black woman who could not break free no matter how hard she tried. I had joined the Army to escape that fate. And here I was, pregnant in a war zone. But my mother was not offering judgment. She was offering grace. She had meant it. She just did not know I already was.

That night, I prayed for Lucky. And I prayed for my mother. Not hopeful prayers. Certain prayers. I knew what God had done for me in that sandstorm. In the blackout of a battle tank. When Strychland tried to rape me and did not succeed. When I opened my mouth to pray and was not gassed or bombed. God had shown up for me again and again in this desert. I knew He would hear my prayers for them too.

Life at Qaisumah was strange. After months of constant duty—guard shifts, burn detail, jump operations—we suddenly stopped. No more assignments. Just waiting.

We were housed in what used to be an airport terminal, a huge room lined wall to wall with cots. One hundred and five soldiers crammed together, men and women trying to carve out slivers of privacy. I made my little space—cot, duffel bag, stack of paperbacks—and stayed quiet.

Boyfriend had returned to Turkey, and Quest was back with Foxtrot Company with the other 11 Bravos. Their absence was a relief. We were a spectacle together—whispers, stares, assumptions. Without them, I could disappear into the background again.

The heat outside was unbearable, so most of us stayed indoors reading. I devoured Stephen King, Mary Higgins Clark, Jackie Collins, and anything that let me escape.

For the first time, we had real bathrooms and showers with doors. After months of makeshift latrines, it felt like luxury. We wore civilian clothes now—T-shirts, shorts, flip-flops. I kept my shirts untucked, hiding the small curve in my belly. Fourteen weeks pregnant. Still praying no one would notice.

The food was terrible. Saudi-prepared meals, mystery meat with rice, endless pita bread and Utz crab chips. To this day, I can't stand the smell of crab chips—it takes me right back to Desert Storm. Sometimes I would eat canned ravioli from a care package and imagine home.

Staff Sergeant Bleach-Bottle Blond and PFC Becky—my personal tormentors—suddenly started acting friendly. They even invited me to the Saudi market once. I went, mostly to prove I was unbothered by them. They spent half an hour trying to buy suntan oil, finally settling on a bottle of cooking oil because the clerk did not understand them. Back at base, they rubbed it on their skin and sunbathed on the blacktop in hundred-degree heat.

I stayed inside and read.

One night, Bleach-Bottle Blond woke up screaming. A Saudi worker had touched her hair while she slept. Within minutes, leadership swarmed around her, comforting, protecting, removing the worker. I watched it all from my cot, thinking how different the response was when danger had my name on it.

The days blurred together after that. Reading, waiting, laying low, praying.

I kept both letters folded in my duffel, wrapped in a clean T-shirt. On the hard days, I would take them out and read them again—once, maybe twice—and remember: I was not forgotten. My life still mattered. God was still in control.

We were all living through separate storms—my mother in DC, me in the Saudi desert—but we were bound by the same prayer. And even before we knew what we needed; God was already answering.

Chapter 24: Back to Germany

Early May 1991.

The last weeks at Qaisumah felt endless. We weren't fighting, training, or preparing anymore. We were just waiting.

The war was over, the missions done, and yet time crawled. Days bled into each other under the same white sun and dry wind. Men smoked. Women read. Conversations repeated themselves. The air carried the weight of deployment fatigue—the weariness of waiting, of being stuck, of a deployment that had overstayed its purpose.

We had survived—but none of us knew what to do with the silence that came after survival.

Sometimes, at night, I would lie awake listening to the hum of the generators and wonder what part of me would make it home. Not the same girl who came here, that much I knew. I was learning how much malice and targeted cruelty from those in charge a person could survive. I didn't see God's grace all around me then, but that understanding would come later with age and wisdom. At that time, I was too young, too wounded, too buried in resentment I couldn't afford to show. Crumbling in front of my tormentors was not an option. There would have been no grace given in the face of my tears. So, I would not dare waste any tears on these malignant leaders.

But even in all that resentment, I knew something. God was working things out for me. All of this was for a purpose I didn't understand. He had protected me in the foxhole. He had carried me through the sandstorm. He had let the war end quickly. He had even orchestrated my pregnancy. I didn't understand His plan, and I didn't think it was my right to know—not yet. I just felt that He was trying to get my attention, and this was the place He knew He could reach me.

Because out here, I had nowhere else to go. My faith was not sentimental anymore. It was worn, bruised, and real.

A song from home drifted through my mind sometimes—Johnny Gill's Fairweather Friend. It reminded me that people could vanish when storms hit, but God never did. The thought did not comfort me in a soft way; it steadied me. I had been through enough to stop confusing comfort with strength.

Then came the orders. We were leaving on May 7, 1991. No formation, no ceremony—just the word spreading through the room like a breeze. Bags packed. Cots folded. Conversations quiet.

When the buses came, nobody said much of anything. We all just moved. After everything that happened, no one had the energy left for grudges or goodbyes.

At the airfield, the desert wind whipped sand against our faces as we boarded the transport plane. I climbed the stairs, boots heavy, heart light in ways I could not explain. I took a window seat, leaned back, and closed my eyes.

When the engines roared and the ground fell away beneath us, the cabin erupted. Clapping. Shouting. Laughter that felt half-relief, half-disbelief.

We had made it. We were finally going home.

For the first time in months, the noise did not grate—it sounded like life again. I leaned my head against the window, smiled, and let the sound wash over me. Hours later, the plane touched down in Germany. The cabin filled with the sound of zippers, shuffling, voices rising with anticipation.

Coming back to Coffey Barracks did not feel like coming home—but it did feel good to be back.

The buses rolled in, and before I could even process the silence of being out of Saudi, we were standing in a gym full of noise and echo. Families cheered, music played, someone had a boom box near the bleachers. Soldiers ran to their people, laughing, crying, lifting children off the ground.

I stood back, letting the surge move around me. Since no one was waiting for me, I left all the fanfare and merriment while it was in full swing. It was deflating—seeing all the married soldiers welcomed back while the single ones stood alone.

Then it was just us again—the same faces, the same chain of command, the same white walls and green duffels. The war was over, but the job wasn't. The air was cooler, wet with spring. I breathed it in like something sacred. For the first time in months, I felt movement that was not forced. Not deliverance in the dramatic sense—just the quiet kind, the kind that sneaks up on you when you realize the worst is behind you and you're still standing.

Germany had always felt good to me. I loved the food, the rhythm, the sense of order here. Coming back now, even with the base in transition, felt like reclaiming a piece of the life I had built before everything unraveled.

Word was that the Base Realignment and Closure—BRAC was starting up again. Our base had been marked for shutdown before Desert Shield, and now that the war was over, those plans were back in motion. Everyone was getting new orders, new assignments, new destinations.

The Army was moving on—and so were we, whether we wanted to or not.

I thought a trip might clear my head. The Morale Welfare Recreation (MWR) office was offering weekend bus tours, and Paris sounded like the opposite of everything I had just survived. Three days, no duties, just sightseeing. I signed up without thinking it through.

It did not take long to realize how unprepared I was for traveling while pregnant. My feet swelled to twice their size somewhere between Germany and the French border. By the time we reached Paris, I could not fit into my slides, and every step felt like walking on fire. I did not complain—I just forced my feet in and tried to keep up with the others.

The hotel was tiny, a narrow box with twin beds pushed against the wall and a shared bathroom down the hall. The guide said that was "normal for Paris." Maybe it was, but it did not feel like a break.

I wanted a nice meal to make up for the rough start to my holiday, but even that backfired. The waiter ignored me until I raised my hand, and when he finally came over, he mocked me—snapped his fingers and barked like I had called a dog. I froze, embarrassed. I did not ask for anything else after that. The food was just… food. Nothing special.

Paris sucked. The people were rude and arrogant.

PARIS

Still, I made the best of it. I walked through a real flea market, took photos, bought souvenirs for my mother—a leather hat and a small black purse. She loved them and kept them for years. I even treated myself to a watch that looked like the numbers had fallen down inside the watch face—"Time Crash," the clerk said. I had never seen anything like it. It was stolen the next month when I went to Turkey, and I've never found another like it.

BERCHTESGADEN

When I got back from Paris, the MWR sent our unit on another trip—this time to the salt mines near the Eagle's Nest in Berchtesgaden. The place was breathtaking, carved out of stone and history, but my mind was somewhere else. I shared a cabin with the only other Black woman in the company, Sergeant White. She had been in Saudi briefly before being sent back, probably for medical reasons. She did not talk much, and neither did I. But her quiet presence felt like peace—no judgment, no side-eye, just space to breathe.

Darnnell at the salt mines in Berchtesgaden

৵

Back at Coffey Barracks, I tried to do something about what had happened in the desert.

First, I went to Staff Sergeant Bleach-Bottle Blond because she was still my immediate supervisor. I told her about Staff Sergeant Strychland. About the assault in the Humvee. About everything. She looked at me with those cold eyes I had come to know so well. "It's going to be your word against his," she said. That was it. No investigation. No concern. Just a dismissal.

So I went to the Equal Opportunity office (EO). I thought maybe—finally—someone outside my chain of command would listen. Someone would help. I sat across from the EO officer and told him the same story. The assault. The retaliation. All of it. He listened, but his expression never changed. When I finished, he leaned back in his chair, unmoved. The message was clear: I was wasting his time. He was not going to help me. No one was. I walked out of that office knowing I was on my own. That nothing was going to change unless I changed it myself. I was so glad I would be leaving Coffey Barracks after all. The BRAC was for the best.

Back on base, I was assigned gym duty. My job was to unlock the weight room, keep it open, and wipe down the equipment when soldiers finished. Most days were long and uneventful. I would get on the stationary bike for a few minutes, trying to stay fit, but my thighs burned and my legs cramped almost immediately. PFC Becky could ride for an hour straight; I would last five minutes and call it quits.

By June, I could feel the difference in my body. My belly was small, but it was there, and my uniform still hid it—mostly. One afternoon, as I left the gym, I passed Staff Sergeant Smith—"Smitty," our supply sergeant. He grinned and called out, "You looking good there, Adams." I kept walking, pretending not to hear.

The truth was, the pregnancy was filling me out in places that got attention for all the wrong reasons. It was awkward, confusing, and I did not want anyone noticing anything at all.

Then one day, the new orders came down. Fort Riley, Kansas. They said it like it was good news. To me, it sounded like exile.

Kansas was a thousand miles from family, from help, from anything that felt familiar. And I was not about to raise a baby alone in the middle of the plains.

So I did what I had learned to do in the desert—take matters into my own hands. I found a phone and called the Department of the Army headquarters in Virginia.

"Personnel Branch, this is Sergeant Collins. How can I help you?" I took a breath. "Hi, Sergeant. My name is Private Darnnell Adams. I'm a 98 Juliet, and I just got orders to Fort Riley, Kansas. I… I need to request a change of station."

There was a pause.

"What's the reason for the request?"

I hadn't prepared a script. I hadn't thought about what I was going to say. So, I just told the truth.

"I'm pregnant, Sergeant. And I don't have any support system in Kansas. My family is on the East Coast, D.C. area. I need to be closer to them. For when the baby comes."

Another pause.

"Are you married, Private?"

"No, Sergeant."

"Father involved?"

"No, Sergeant."

I heard papers shuffle.

"Okay," she said. "Where do you want to go?"

"Fort Meade, Maryland," I said immediately. "It's close to D.C. Close to my family."

More shuffling.

"Fort Meade… let me see what we have."

I held my breath. Please, God. Please.

"Alright," she said finally. "I can get you to Fort Meade. You'll be working at the National Security Agency, NSA. That work for you?"

"Yes, Sergeant. Thank you. Thank you so much."

"Your new orders will be in the system in a few days. Check with your admin office."

"I will. Thank you, Sergeant."

"Good luck, Private."

And she hung up.

I stood there holding the phone for a long time after the call ended.

I had done it.

I was going to Fort Meade—going home.

Well, close enough.

When the paperwork came through, my leadership tried to act unimpressed, but I could see the surprise in their faces. They'd been eager to tell me I was headed to Kansas. I took quiet delight in telling them I was not.

"Fort Meade," I said, handing over the orders. "Home." It was the first time since joining the Army that I felt like the system had worked in my favor.

※

I hadn't seen Sonny Quest since the desert—since that night that started everything and ended everything at the same time.

Weeks had passed. The war had ended. We were all back at Coffey Barracks, pretending things were normal. But the moment I saw him again, walking toward me through the main gate, it all came rushing back.

He might've been on duty or just passing through, but his face said everything—tight jaw, eyes forward, no warmth, no recognition. He walked past me like I was invisible.

I turned toward him. "Sonny."

He did not answer. Did not even slow down.

"I'm sorry," I said quietly.

Nothing. Just the sound of boots on pavement.

I took a breath. "I'm pregnant," I said. "You're the father."

That stopped him. He froze, then finally looked at me. "Is it true?"

"Yes," I said. "But I'm not trying to broadcast it. If someone decides to be petty, they could ground me from flying home because of how far along I am—airlines and even the Army had rules about that."

And when I said "someone," I meant my own unit. They'd been showing me who they were all along.

He stared at me for a moment, expression unreadable—not angered or in disbelief, just emotionless with a look of tired disappointment. Then he nodded once, almost to himself, and walked away.

I should've known better. He'd already been acting like a man who could not handle rejection quietly. I just did not know yet how long he'd hold on to it.

For a brief moment, I thought that meant he understood. That he'd keep it between us.

He did not.

A few hours later, one of the soldiers from my company pulled me aside and asked if it was true—what Quest had said about me being pregnant.

I froze. The words hadn't even cooled, and already they'd spread.

I denied it without hesitation. Not because it was not true, but because I could not risk being grounded, not this close to getting home.

The lie burned, but the truth would've trapped me here—with them.

By the end of the day, I had learned two things for certain:

Silence had kept me alive this long.

And Sonny Quest was not the ally I had once thought he was.

⁊♦

Chapter 24: Back to Germany

Before leaving Coffey Barracks, I spent some time with my friends Mike and Christine Heflin.

We had gone through AIT together back at Fort Devens, Massachusetts, and they'd always been good to me.

They were living off-post now, married, with a new baby girl named Kelsey who'd been born that spring while we were still in Saudi.

I visited them a few times at their apartment. Christine was glowing, mother-tired but happy, and Mike was just as gentle as I remembered. They welcomed me in without questions or judgment, and being around them felt easy—normal—something I hadn't felt in a long time.

I don't think I ever told them the truth about the pregnancy.

Maybe I let them assume it was Boyfriend's, maybe I just stayed quiet. It was all too raw, too complicated to explain. But being in their warm little apartment, holding their baby, eating Christine's home cooking, reminded me what kindness felt like. It was the first time since the desert that I could sit still and feel safe.

Darnnell at Mike and Christine's

Then I made a decision that still makes me shake my head. I requested thirty days of leave—June 20 to July 20, 1991—to go to Turkey.

I had written my mother in May, telling her I was going. I found the old letter recently—dated May 22, 1991. But I don't think I told anyone else at Coffey Barracks where I was headed or why.

So, I was going to Turkey.

Dear Ma & Char— 22 ~~June~~ May 91

I went to Paris for three days and I picked up some souvenirs for both of you. Hope you like them. Sorry there were no "Lucky" sizes but I did get her some cute little thing at the PX.

Good news. I called DA (department of the Army) and my next unit assignment will be Ft. Meade Md. Not far from home.

I should be there no later than October. That's my reporting date.

Hope everything's OK there. I miss you both.

Paris sucks. The people are rude & arrogant. The city of Paris' basically looks like downtown D.C., except that its bigger.

I'm going on 30 day leave on 20 June to 20 July to Turkey. ▓▓ and I will probably travel during that time also. We're not going anywhere near the mid east though. Well, I hope you like these little gifts.

Chapter 25: Journey to Sinop

June 1991. Ramstein Airbase Germany.

Once I got back to Coffey Barracks from Saudi Arabia, the guilt was eating me alive.

I had crushed Boyfriend in the desert. Made a fool of him in front of everyone. He'd traveled to a warzone to see me. He did not have to, but he came anyway. He risked everything because he loved me.

And I betrayed him.

He'd left Saudi earlier than I had, returned to his assignment at Sinop, Turkey, and I could not stop thinking about what I had done to him. It was not enough to let it go. I could not just move on and pretend it hadn't happened.

He deserved a face-to-face apology. A real one. The kind you can't give in a letter or over a phone call.

So after the Paris trip, I made arrangements to go to Turkey. I requested thirty days of leave. My command approved it without asking too many questions. I had told my mother I was going to Turkey when I wrote to her in May. But I don't think I told my friends at Coffey Barracks. I'm pretty sure I told Boyfriend I was coming, and I'm pretty sure he did not want me to. But I was going anyway.

I had no cell phone. No way for anyone to reach me once I left Germany. I was pregnant, and about to travel alone through Turkey to apologize to someone who had every right to never speak to me again.

Looking back now, I can see how reckless it was. How foolish. How dangerous. But at the time, all I could think about was making it right.

Maybe joining the Army had made me bolder than I used to be. I had been making bold moves since before basic training—leaving home, surviving the dehumanization, deploying to war. This felt like just another mission. Another impossible thing I had to do.

God and I were getting closer every day. I prayed constantly. For protection. For guidance. For courage.

I was going to need all three.

❦

I flew out of Ramstein Air Base on a Military Airlift Command (MAC) flight. Because I was active duty, I was prioritized ahead of civilians and retirees. I don't remember who else was on the flight. I was not paying attention. I was focused on the journey ahead, on what I was going to say when I finally saw Boyfriend again.

The plane was a C-130—a huge military cargo aircraft. I had never been on a military plane before. I did not know what to expect.

When I stepped inside, I was stunned. It was enormous. Almost cavernous. Nothing like a commercial airplane. The interior was raw, utilitarian, stripped down to function. No windows along the sides where passengers sat. Just metal and rivets and the hum of massive engines.

I was strapped into a web-type seat—mesh netting stretched across a metal frame, nothing like the cushioned seats I was used to. It was uncomfortable, strange, and a little unnerving.

They handed out box lunches—basic sandwiches, ham and cheese or boloney and cheese, and an apple. They also gave us orange earplugs because it was incredibly loud inside that metal plane. The engines roared, echoing off the walls.

I sat there with my earplugs in, sandwich untouched, looking around at the other passengers, at the cargo secured in the center of the plane, and thought: What am I doing?

But it was too late to turn back now.

The flight to Incirlik Air Base in Turkey felt endless. Hours in the air, strapped into that web seat, my mind racing with everything I needed to say.

Incirlik Air Base was a major U.S. military hub in Turkey, especially after the Gulf War. It was huge, busy, full of personnel coming and going for Operation Provide Comfort and other missions in the region.

But I was not staying at Incirlik.

Boyfriend was stationed at Sinop—a small, isolated town on the northern coast of Turkey, on the Black Sea. It was hours away from Incirlik. There were no direct military flights to Sinop. The airport there was too small, too remote.

So I had to take connecting flights and ground transport.

From Incirlik, I made my way to Ankara, Turkey's capital. I don't remember exactly how I got there—maybe a military transport, maybe a domestic flight. Everything about that journey is a blur now.

What I do remember is arriving in Ankara exhausted, overwhelmed, and completely unprepared for what came next.

The bus to Sinop did not leave until the next morning.

I had to spend the night in Ankara, but I had nowhere to stay. I don't remember how it happened. I don't remember who they were— whether it was a man or a woman, Turkish or American, military or civilian. But someone took pity on me. A stranger.

Maybe someone I stopped to ask for directions saw my desperation, knew I was a stranded foreigner in need of help, and God put it in their heart to offer me a safe place to rest. God knows I had prayed for help and I believe he sent them to me. But, before I went with them, I prayed for God's protection. And somehow, I could tell they were safe. They must have proven to me they weren't dangerous, because I know I wouldn't have gone otherwise.

But I needed to trust someone. And they made me feel like I could trust them.

I slept in their room, or their home—I don't remember the details. Just that I was grateful. Exhausted. And still praying for God's protection over every step of this journey.

When I woke up the next morning, I thanked them and left for the bus station. I never saw them again.

The bus to Sinop left early in the morning. I had my luggage— probably a duffel bag, maybe a backpack. I don't even remember what I packed. I was so unprepared. So naïve about the journey ahead of me.

I got on the bus and found a seat. And then we started climbing into the mountains.

The route from Ankara to Sinop cut through the Pontic Mountains, a rugged, forested region that rose steeply toward the Black Sea coast. The roads were narrow, winding, carved into the sides of mountains with sharp curves and steep inclines.

There were no guardrails.

The bus was as wide as the road.

And the drop-offs—the sheer cliffs that fell away into the valleys below—were right there. Right outside my window.

I looked out to my left and saw the edge. Saw how close we were. Saw how far the valley dropped.

My stomach lurched.

The driver navigated with surgical precision—every turn calculated, every movement deliberate. He was skilled. Confident. He'd driven this route a thousand times.

But I was terrified.

The road wound higher and higher. Steep grades. Sharp curves that seemed to come out of nowhere. The bus leaned into the turns, hugging the cliff edge as we climbed.

I had never experienced anything like this. The exposure. The heights. The way the bus seemed to balance on the edge of nothing.

I was exhausted. From the C-130. From sleeping in a stranger's house. From carrying all this guilt and fear and hope across continents just to apologize to someone I had hurt.

I said a prayer. God, protect me. Keep me. Let me get to my destination safely. And then, somehow, I fell asleep.

I shouldn't have been able to sleep. Not with the cliff edge right there. Not with the curves and the heights and the terrifying exposure.

But I had done all I could do. I had made my choice. I had gotten on the bus. I had come this far.

The rest was up to God.

I woke up as the bus began its descent toward the coast.

The Black Sea came into view—dark blue, stretching endlessly toward the horizon. And there, on a peninsula jutting out into the water, was Sinop.

Beautiful. Isolated. Remote.

I had made it.

Boyfriend was no longer my boyfriend. I knew that. He knew that. But when I arrived in Sinop and finally found him, he treated me kindly. He was still hurt. I could see it in his face, hear it in his voice. What I had done to him in Saudi had crushed him. Made him feel like a fool in front of everyone. But he did not turn me away.

I stayed at a local Turkish inn—not fancy, just a small, simple place with a bed and a bathroom. The maids were all short women who did not wear shoes. They had the fattest, shortest feet I had ever seen, and I remember thinking how different everything was here. How far I was from anything familiar.

One day, I took off my "Time Crash" watch—the one I bought in Paris—and left it on the nightstand. When I came back, it was gone. I was furious. I tried to interrogate the cleaning ladies, but they did not speak English, and I did not speak Turkish. I gestured. I pointed. I tried to make them understand. But it was useless. The watch was gone, and I never got it back.

Every day, Boyfriend showed me around Sinop. He took me to this wonderful pizza place called Sbarnak. The pizza was delicious—better than I expected in a small Turkish town on the Black Sea.

We went shopping. I bought an authentic Turkish robe, some wooden Coffey table boxes, and a puzzle ring. Little souvenirs to take back with me. Proof that I had been here, that this trip had been real.

He even took me to see where he worked—the command center at Sinop. He had lucked up and got a strategic 98 Juliet assignment. We both had trained for this kind of intelligence work. I was proud of him. Glad he'd landed somewhere good.

We talked. We cried. We laughed. We ate.

He showed me a very nice time, but we were not intimate at all. This was not about romance or rekindling anything. It was strictly about giving him the courtesy of a true and honest apology.

One afternoon, sitting at a café overlooking the water, I finally said what I had come all this way to say.

"I am so sorry," I told him. "For everything. For what I did in Saudi. For how I hurt you. It was my fuckup. Nothing you did. You didn't deserve any of it."

He looked at me for a long time, not saying anything. "I know you're still hurt," I continued. "And I don't expect you to forgive me right away. But I needed you to know—I needed to tell you in person—that it was all my fault. And I'm so, so sorry."

He nodded slowly. "I appreciate you coming all this way to say that."

"I wish things had been different," I said. "I wish I hadn't messed everything up."

"Me too," he said quietly.

We sat there in silence for a while, watching the waves roll in.

"I want us to end as friends," I said finally. "I won't contact you after this. I won't bother you. But I needed you to know the truth. And I needed to say I'm sorry."

He looked at me and nodded. "Okay."

And that was it.

We finished our Coffey, and we moved on.

That night, alone in my room at the inn, I hit my lowest point.

I had done what I came to do. I had apologized. I had faced him. I had tried to make it right so he was not irrevocably scorned by women for the rest of his life.

But now it was over. Really over, and I wished Quest could have handled things the way Boyfriend did. Comparing the two men was not about picking sides—it was about finally seeing who they truly were, without the fog of loneliness or war.

It was hard to believe Quest was the same man who once carried my burdens quietly, who saw my suffering before anyone else did. He didn't have to forgive me—I understood that. But he couldn't even talk to me human to human for the sake of his unborn child. The distance he'd created in the desert never closed—not even now.

166

Seeing him again only confirmed we were done. He was curt, sullen, and moody. All the warmth and connection between us—gone. Just like that. And that was revealing about who he truly was as a man.

And then he told people. Maybe asking him to keep it quiet was a lot—I can see that now. But the real issue wasn't that he told. It's that he couldn't even have a civil conversation with me about the child we had created together. He shut me out completely. That wasn't about keeping secrets. That was about character.

If he'd truly loved me, there would have been some remnant of that love left—some kindness, some understanding. But there was not.

Boyfriend, even in his hurt, had shown me grace. He'd been kind. He'd accepted my apology like a man—treated me with respect even when he didn't have to.

Boyfriend and I were done. Quest and I were done. I was pregnant, alone, thousands of miles from home, with no idea what was next in my future.

For a moment—just a brief, terrifying moment—I felt completely alone. Like I was the only person on planet earth. Like I had no one.

Maybe it was pregnancy hormones. Maybe it was exhaustion. Maybe it was the weight of everything I had been carrying finally crashing down on me all at once. But somehow, I snapped out of it. I remembered: God loves me. My family loves me. And I love me. I was sad over a dumb mistake. A very public, very messy mistake. But it was not that serious. I did not kill anybody. This was not life-altering. Well, except for the baby. But people get pregnant and have babies every day. It's life. I was going to be okay.

∂❧

Before I left Sinop, I said goodbye to Boyfriend one last time. "I wish you peace and happiness," I told him. And I meant it. He smiled—a small, sad smile—and nodded. And then I left. I never saw him again. Never reached out. Never contacted him. That chapter of my life was closed.

The journey back to Germany was long, exhausting, and uneventful. But something had shifted. The guilt I had been carrying—the weight of what I had done to Boyfriend—was gone. I had faced him. I had apologized. I had done the work of truly repenting for the pain I had caused. And now I could move forward.

When I got back to Coffey Barracks, I felt lighter. Clearer. Ready. Ready to get my orders. Ready to move to Fort Meade. Ready to focus on the baby. Ready to start the next chapter of my life.

Looking back now, I can see how foolish that trip was. Traveling alone through Turkey. Pregnant. No cell phone. Sleeping in a stranger's house in Ankara. Riding that terrifying bus through the mountains. Trusting that God would keep me safe. It was reckless. Dangerous. The kind of thing you do when you're young and naïve and convinced that doing the right thing is worth any risk. My mother is going to be furious when she reads this. My daughter too. And lawd-have-mercy, if Deidra ever did something this reckless—traveled alone through a foreign country while pregnant, stayed with strangers, rode death-defying buses through mountains—I would have a heart attack. I would lose my mind.

But I'm still here. God kept me. He protected me every step of that journey. And I learned something important: sometimes, doing the right thing means taking a leap of faith—even when it's scary, even when it's foolish, even when no one else would understand.

Seeing things now from my 54-year-old eyes, I believe God wanted me to make things right with Boyfriend because I had been especially thoughtless with his heart. I think it was part of my humbling—part of the heartbreaking that was necessary for God to rebuild me into the best version of me that He was making.

Chapter 25: Journey to Sinop

That trip—that crazy, reckless, beautiful trip—was one of the most foolish things I've ever done. But it was also one of the bravest.

And I wouldn't take it back.

Chapter 26: Fort Meade and Deidra's Birth

September 1991.

I arrived at Fort Meade on a warm, humid afternoon. The air felt thick, different from the dry desert heat, different from the crisp German air. It felt like home. Or close to it.

Fort Meade was only about twenty-five minutes from DC. A quick drive up the Baltimore-Washington Parkway. Close enough that I could see my family on weekends. Close enough that I was not completely alone anymore. And for the first time in months, I felt something that might've been relief.

The barracks at Fort Meade were different from Germany. They were apartment-style, small units with a bedroom, a bathroom, and full shared kitchens on each floor. Private. Clean. Modern. Everything the barracks in Germany hadn't been. And I had one all to myself. No roommate. No shared bathroom. No one watching me. Just me.

I was six months pregnant when I arrived. My family and my new unit already knew. There were no more secrets left to keep, only time to fill before the baby came. I went to Clothing Sales and bought maternity BDUs—the official ones with the stretch waistband and the extra-room jacket. It felt strange at first, buttoning up a uniform made for pregnancy, but it also felt right. I was still a soldier, still showing up. I wore them loose, kept my head down, and tried to focus on staying healthy. Everything had already changed—I just hadn't slowed down long enough to feel it yet.

I reported to my new unit and started in-processing. Medical appointments. Briefings. Paperwork. All the usual stuff.

At my first medical appointment at Fort Meade, I finally sat across from a doctor instead of a field medic. "I'm pregnant," I told her. She was a middle-aged Black woman with kind eyes who carried herself like she had seen everything before. She did not look surprised—just nodded and made a note in my chart.

"How far along?" "About six months."

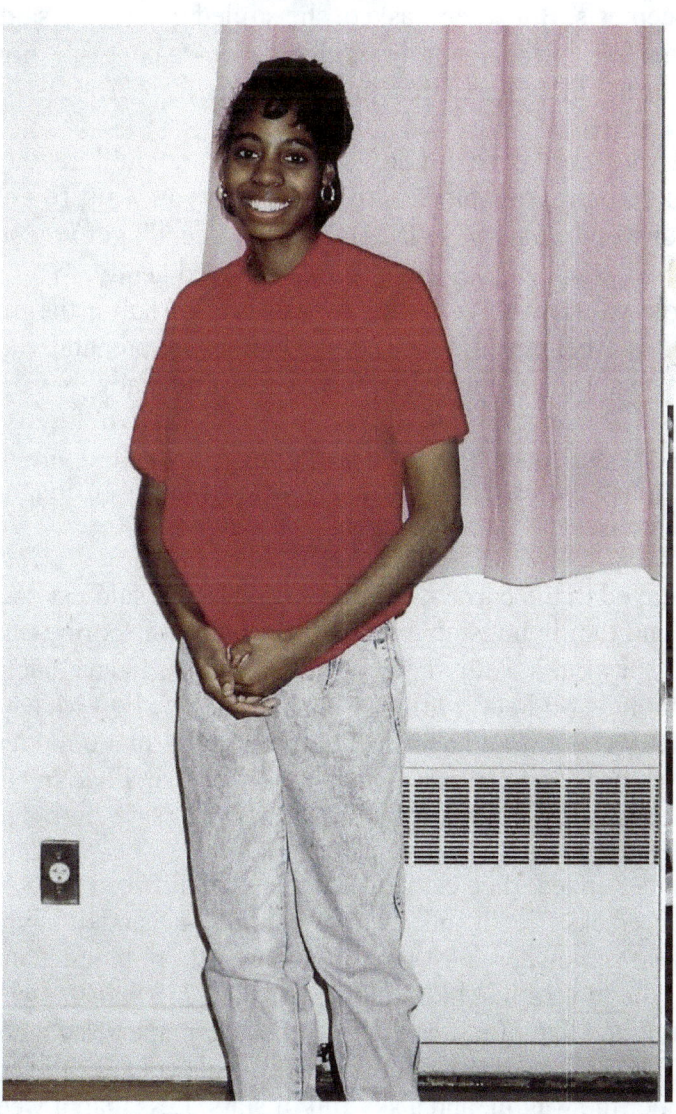

"Have you had any prenatal care?" "No, ma'am," I said. "I was deployed." She paused for a long moment, pen still in hand, then nodded again. "Alright. We'll start fresh and make sure everything's on track."

As she ran through the checklist, I found myself wondering if she had mentioned that test Christine had told me about—the amniocentesis. But when I asked, she smiled gently and said, "That's done earlier, in the second trimester. You're past that now." Her tone was not judgmental, just matter-of-fact.

"We'll take good care of you," she said. "And you'll need to start thinking about your birth plan. Where do you want to deliver?" "Walter Reed," I said. "In DC." "Good choice. I'll get you set up with their OB clinic." She paused, then added gently, "You're going to be okay, Private. We'll take care of you." And for the first time in months, I believed it. She sent me home with prenatal vitamins—the first real thing I could do to take care of my baby.

The pregnancy became official after that. My chain of command was notified. My duties were adjusted—no more heavy lifting, no more field exercises.

But I stayed in the barracks. In the Army, single soldiers could not move into family housing until after the baby was born—until there was, as they put it, "a living dependent." As morbid as that sounded, it was policy. So I stayed right where I was. The Fort Meade barracks were new and apartment-style—clean, quiet, modern. I had my own space and privacy, which was more than I had in Germany. And honestly, I did not mind. I was fine there. Safe enough.

September ended, then came October. On Halloween night, I let one of my barracks friends talk me into going on a haunted hayride out on a farm somewhere past Laurel. She was short, blonde, and wide-eyed—she looked just like Ralphie from *A Christmas Story*, and even laughed like him. I can't remember her name now, but she was kind. The ride was gentle, nothing wild—just cold air, the creak of hay under the wagon, and a sky full of stars. Less than a week later, I went into labor.

On Wednesday, I woke up at 1 AM with a stomachache. Because I was asleep and groggy, I just turned to the other side, hoping it would go away. So I turned over and tried to go back to sleep. But the pain came back. And again. And again. Each time stronger than the last.

It took me three rounds of waking up before I realized: These weren't stomachaches. These were contractions. The baby was coming.

I sat up in bed, heart pounding. *Okay. Okay. Don't panic. You knew this was coming. You're prepared.* Except I was not prepared. Not really.

I got dressed, slowly, carefully, stopping every few minutes when a contraction hit. I grabbed my hospital bag, packed weeks ago, just in case. And I knocked on the same girl's door who reminded me of Ralphie. Her name was... God, I can't even remember her name now. She had agreed to be my ride to Walter Reed when the time came. And the time was now.

She answered the door in her pajamas, squinting at me. "Darnnell?" "It's time," I said. "The baby's coming." Her eyes went wide. "Now?" "Now." "Okay. Okay. Let me get dressed." She scrambled back into her room, and I stood in the hallway, breathing through another contraction.

We got into her car, a small hatchback, and started driving. The roads were empty at 2 AM, but they were also icy. It had been cold lately, and black ice was a real threat. My friend drove carefully, slowly, and I gripped the door handle every time we hit a bump or pothole. "Sorry!" she kept saying. "I'm trying!" "It's okay," I said through gritted teeth. "Just... get me there."

೨ல

We made it to Walter Reed around 3 AM. I checked in at the front desk, gave them my information, and they took me back to triage. An OB doctor on duty examined me. "You're not dilated yet," she said. "What?" "Your cervix hasn't dilated. You're not in active labor yet." "But I'm having contractions." "I know. But you're not ready to deliver. Come back tomorrow."

I stared at her. "I can't come back tomorrow. I'm here now. And my ride already left." The doctor sighed. "Let me talk to the admit team." They admitted me. Not because I was ready to deliver, but because I had no way to get home.

They put me in a room and told me to take a walk. "Walk around the hospital. It'll help move things along." So I walked. Up and down the hallways. Through the waiting rooms. Around the maternity ward. For hours.

My mom showed up around 8 AM. I had called her after arriving. So she came. And she stayed. "You're going to be okay, baby," she said, holding my hand. And I wanted to believe her.

At some point, they offered me an epidural. I took it. And for a few hours, the pain eased. But then it came back. On Wednesday afternoon, the nurses gave me Pitocin to speed up the labor. And suddenly, the contractions were unbearable. I thought the contractions were bad before. But this was different. This was pain that took over my entire body. Pain that made me forget how to breathe. Pain that made me understand why women screamed during childbirth.

I labored all day and all night. The epidural had since worn off. And I was too weak to push anymore. Twenty-four hours of contractions. Of pushing. Of exhaustion so deep I could not think straight. Too tired. Too done.

"I can't," I said to the doctor. "I can't do this anymore." "Yes, you can," she said. She was a young Black woman, maybe not much older than me. "You're almost there. One more push." "I can't." "You can. Come on, Darnnell. Your baby wants to meet you."

I pushed one more time. With everything I had left. And at 1:48 AM, she was born.

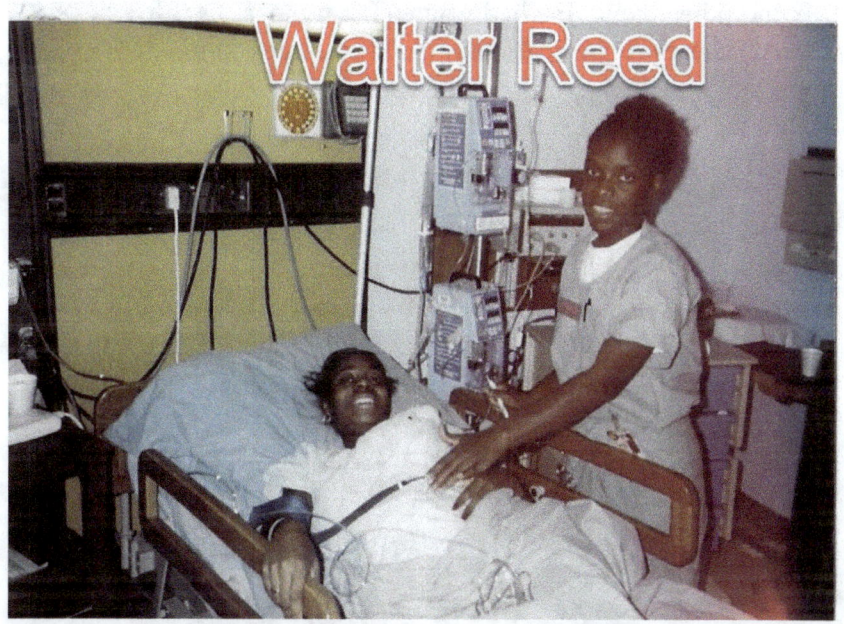

Deidra Darnnell Adams. Seven pounds, thirteen ounces. Perfect. The doctor held her up, and she let out a cry. The loudest cry I had ever heard. "She's got the most well-developed lungs of any baby I've ever heard," the doctor said, laughing. And I started laughing too. Probably from tired delirium. From relief that the hard part was over. From joy that God brought me through this. I had survived. We had survived.

୨෴

They cleaned her up and handed her to me. And I looked down at her tiny face, eyes squinted shut, mouth open in mid-cry, and I fell in love. Completely. Instantly. Irrevocably.

"Hi, baby girl," I whispered. "I'm your mama." She stopped crying. Just for a second. Like she recognized my voice. And I thought: *This is why. This is why I survived. For her.*

My mom was smiling happily too. "She's beautiful, Darnnell. She's perfect." "I know." "What are you going to name her?" "Deidra," I said. "Deidra Darnnell."

My mom hesitated. She had wanted me to name her after her—Cheryl Marie—and she did not hide her disappointment. But this time, I did not bend. "Your name," she said finally, still sounding a little salty. "Yeah," I said, smiling down at my daughter. "So she'll always know she's mine."

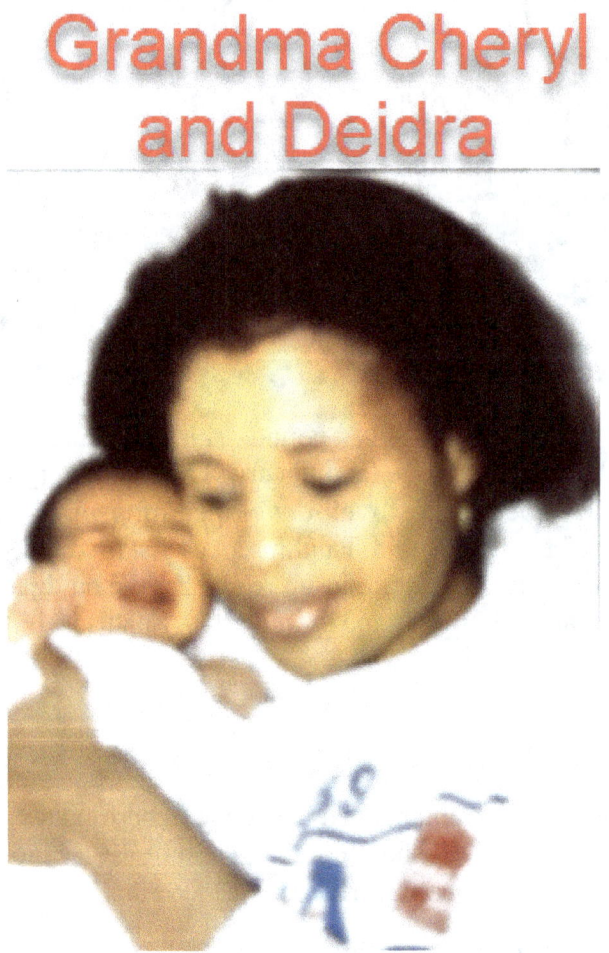

Grandma Cheryl and Deidra

They took Deidra to the nursery so I could rest. And I fell asleep almost immediately. Twenty-four hours of labor. I had earned it.

When I woke up, the first thing I did was ask for her. The nurse brought her to me, and I held her again. Studied her. Her tiny fingers. Her dark hair. Her perfect little nose. She was real. She was here. She was mine.

I prayed over her. "God, thank You. Thank You for keeping her safe. Thank You for getting us through. Please protect her. Please give her everything I never had. Please let her grow up strong and loved and safe. Please." And I kissed her forehead. And I promised her: "I'm going to give you the best life I can. I promise."

We did not go home right away. Deidra developed jaundice—her tiny skin turned the faintest shade of yellow—and she had to stay under the ultraviolet lights for treatment. They kept her in a little bassinet with a soft blue glow, wearing only a diaper and tiny eye covers. I sat beside her for hours, watching that gentle light wash over her, praying it would work.

The nurses were kind, but those days were hard. I tried to breastfeed, but nothing came. My breasts just wouldn't cooperate, and the baby was not getting any milk. The nurses showed me how to knead my breasts, to "wake them up," as they said, but it barely helped at first. They explained that sometimes milk took a few extra days to come in after a long labor—something called delayed lactation—but that did not make it any easier to watch my baby cry from hunger.

So they gave Deidra formula to make sure she was getting enough nourishment until my milk finally came in. I remember feeling both grateful and guilty—grateful that she was being fed, guilty that my body could not do it on its own yet. We stayed a week before she was finally cleared to go home. When the doctor said she was stable, I exhaled for what felt like the first time in days.

When we were discharged, my aunt Teresa offered to let us stay with her. She and my cousin Edna had a vacant finished basement in their house in Montgomery County that we could use until my military quarters were ready. I thankfully accepted.

It was a little cold down there, but it was warm in every way that mattered. The space was cozy, quiet, and safe. There was an orange high-back sofa chair that became my favorite spot—I would sit there for hours holding Deidra, just watching her sleep or studying her little face. I did not want to put her down.

I remember one afternoon holding Deidra out as an offering to God, asking Him for divine protection over her life—that He would personally guard her every day of her life. I prayed that He would provide above and beyond what even two loving earthly parents could do. That He would be her eternal Father, to love and protect her in ways I could not yet imagine. And I promised that she would know Him—that I would raise her to know, love, and revere Him.

When my younger cousin Edna came home from school, she would come downstairs to visit. We would take turns holding Deidra and she would tell me about her day at school, her friends, or whatever was on her mind. Those talks meant a lot to me. Life finally felt calm.

It was peaceful not having to work or think about work for a while. But my maternity recovery time was only twelve weeks, and it was clear to me that it would end all too soon.

Before I went back to work, I made the most of every day. I would bundle Deidra in her little carrier and took her on walks around the neighborhood. The air felt crisp, the world felt gentle again, and I remember thinking: *This is joy.*

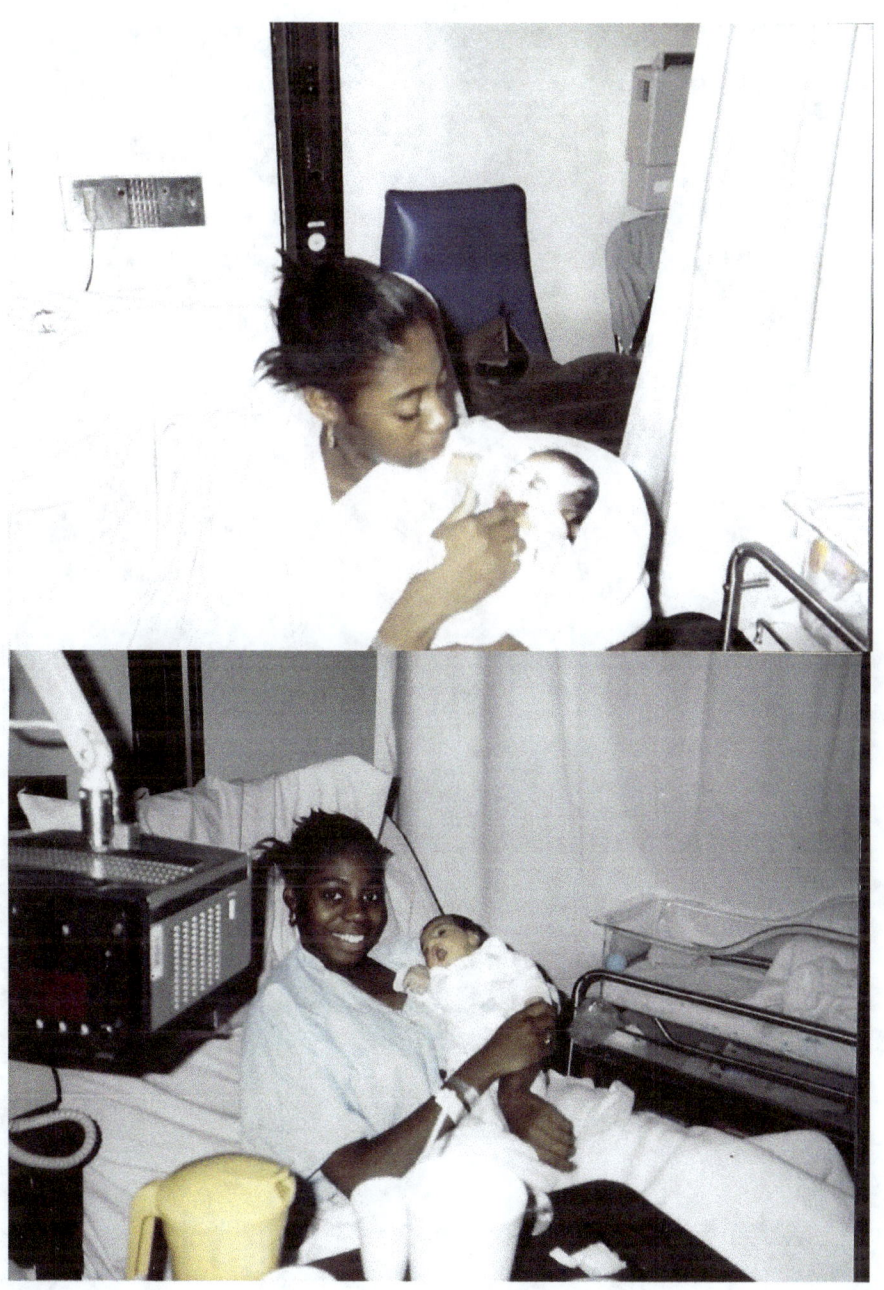

Chapter 27: SSG Busch — My Mentor, My Friend

When I first got to Headquarters and Operations Company (HOC), 742nd Military Intelligence Battalion at Fort Meade, before Deidra was born, I had been assigned to a new team.

A new team leader.

Staff Sergeant Busch.

And for the first time since joining the Army, my team leader was a Black woman.

Staff Sergeant Busch was from the Tidewater area of Virginia — Norfolk, I think. She was smart, confident, and no-nonsense. She was married to a German man, which I found interesting because I had just come from Germany and hadn't seen many interracial couples there.

They had a son together, Demarkus, who was maybe a year old when I met her. Light-skinned. Adorable. Rambunctious.

I liked Staff Sergeant Busch immediately. Not because she was Black — though that helped — but because she treated me like a person. Not a shiftless soldier. Not a worthless Black girl from the inner city, the way my previous white leadership had made me feel. They treated women like me as if we were wild horses that needed to be broken before we could be trusted. But Staff Sergeant Busch treated me like a human being first and foremost— because she was a Black woman too. It was a refreshing change of pace.

We bonded over Germany. Swapped stories about the food, the travel, the culture. She told me about her husband, about how they'd met, about how German men loved Black women.

I did not have the same experience, but I listened. And I was happy for her. I visited their house a few times before Deidra was born — a neat, cozy on-base home at Fort Meade. Her husband was quiet. Reserved. He'd nod hello and then disappear into another room, leaving us to talk. I remember thinking their relationship seemed... tense. But I did not say anything. It was not my business.

Most evenings when I stopped by, Staff Sergeant Busch would be in the kitchen cooking baked smothered turkey wings over rice and gravy with a vegetable on the side. The smell filled the house and made me feel at peace. It felt good to sit at a real table, to laugh, to be around a family. It reminded me of Mike and Christine back in Germany — people who took me in and made me feel human again.

One evening, while waiting for dinner, I was holding her son on my lap. He was a lively little boy with a hard head and a fearless streak. He was wiggling on my knees when he suddenly leaned forward, then snapped his head back — square into my front teeth. The pain shot straight to my brain. For a second, I thought my teeth were gone. I sat there holding my mouth, trying not to cry or cuss. When I finally looked, there was no blood, no missing teeth — just a throbbing ache and a lesson learned about babysitting busy toddlers. I laughed later, but in the moment I was sure I would be spitting out chicklets.

After Deidra was born, Staff Sergeant Busch got pregnant again. When she had her second son, he looked completely different from Demarkus — darker, more like her.
Apparently, her husband questioned whether the baby was his.

She told me about it one day when I visited.

"Can you believe it?" she said, bouncing the baby on her hip. "He actually asked me if this one was his. Like I would cheat on him."

I did not know what to say.

"That's... rough."

"Yeah," she said. "But he came around. He just needed to see the baby's features come through."

But again, I did not say anything. After my desert mess, I was in no position to comment.

We stayed friendly for a while. After I started working at NSA and picking Deidra up after work, I could not make our frequent dinner gatherings anymore. Life just got busier. We were both mothers, both tired, both trying to keep up. But I really enjoyed spending time with her. She was like a mentor, big sister, and mom all at the same time. Staff Sergeant Busch also guided other young female soldiers in our unit. It seemed like all of us were trying to figure out who we were, and we truly appreciated her gentle wisdom and steady guidance.

She was also a mentor and close friend to another young soldier, Sara, who was a Specialist like me. We weren't close, but years later I ran into her again at Fort Drum, New York — under the hardest of circumstances. She had a baby girl who was very sick, and I could see the weight of it all on her face.

I remember thinking how much she could have used Staff Sergeant Busch's steady voice in that season.

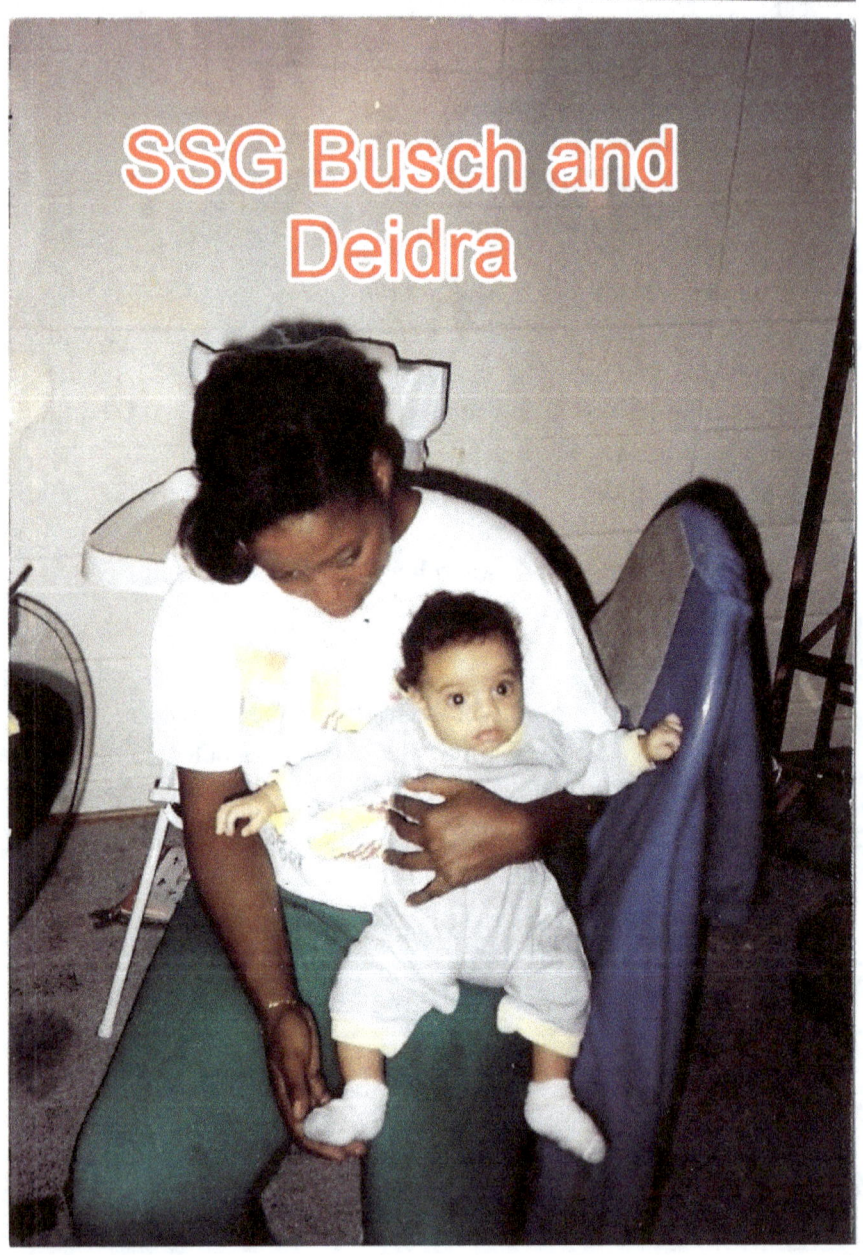

Chapter 28: The Matchbox and Dana

December 1991

After a month in Aunt Teresa's basement, it was time to stand on my own again. The convalescence had done its job — I had healed, found my footing as a new mother, and was ready to face whatever came next.

When the call came that my military housing was ready, I packed our few things into the car and drove back to Fort Meade.

The address read: 1923 Reece Road, Apartment F.

The building looked old — like something the Army had forgotten to tear down. One bedroom, one bath, a small kitchen, creaky floors, and radiator heat that worked when it felt like it. But it was ours, and that was enough.

I put Disney stickers on the wall near Deidra's crib, trying to make the space cheerful and alive. She was such an easy-going baby, but I still wanted to impress her. She's always had high standards —still does.

The apartment was small and the walls were thin. I could hear every footstep and conversation from next door, but I refused to see it as depressing. It was our fresh start, and I wanted it to feel that way.

So I made the best of it — hung curtains, bought a little lamp, and filled Deidra's corner with soft blankets, stuffed animals, and anything that made the space feel cozy and loved. When the lamp was on and she was asleep, the whole apartment glowed. It was not much, but it felt like peace.

Darnnell and Deidra
on Reece Road

Because even if I could not give her everything, I could give her that.

Going back to work was hard. I had taken twelve weeks of maternity leave — the standard amount — but it was not enough. Deidra was still so small. Still nursing. Still needing me every few hours. I had to find someone I could trust.

Most babysitters did not feel right — too expensive, too far, too something. And then I met Dana. Dana Gedowski was a Polish woman in her forties with kind eyes, a warm smile, and short blond hair. She reminded me of the Polish mother who'd lived across the street from my grandmother's house on Jackson Street — quiet, steady, the kind of woman who radiated care.

She came to my apartment for the interview and immediately asked to hold Deidra. I handed her over, watching closely. Dana spoke softly to her in Polish, and Deidra studied her face for a moment — then smiled. Not a gas smile. Not a reflex. A real one.

"She likes you," I said. Dana smiled. "I like her too. She's beautiful." "You must be sad to leave her and go back to work, yes?" She knew exactly how I felt. "Yeah," I said quietly. "I'm not looking forward to it." "I take good care of her," Dana said. "You don't worry. I have three children. I know babies." And I believed her.

For the first few weeks, Dana came to my apartment every morning at six and stayed until I returned from work around five. She fed Deidra, changed her, played with her, and every evening when I came home, my baby was clean, fed, and happy.

After a month, Dana asked if she could keep Deidra at her own house instead.

"It's easier for me," she said. "More space. More toys. My husband loves babies — he would love to meet her." I hesitated. Taking Deidra out of my sight felt scary. But I trusted Dana, and Deidra adored her. So I said yes.

Dana's house was nearby in Severn, just a few minutes from Fort Meade. A small, neat home with a fenced yard, flowers in the garden, and the smell of home-cooked food inside. It felt like a grandmother's house — safe, warm, and full of love.

Every morning, I would drop Deidra off before work and pick her up afterward. Dana always had her ready, fed, changed, smiling. She would tell me about their day — "She smiled at the dog," "She held my finger so tight," "She's growing so fast."

I would listen, grateful beyond words. Dana was not just watching my daughter — she was loving her.

I was still breastfeeding, nursing Deidra each morning before work and pumping during lunch to bring milk home in a cooler. It was exhausting, but it mattered to me. I wanted her to have the best start possible — to know, even when I was not there, that I was taking care of her. By three months, Deidra was eating more than just breast milk. I started adding cereal to her bottles because she was always hungry. Always. She would finish a bottle and cry for more, and I would laugh.

"You're going to eat me out of house and home, aren't you?"

But I did not mind. She was healthy, growing, thriving — and that was all that mattered.

Those first few months were the happiest I had been in years. Yes, I was exhausted. Yes, I was broke. Yes, I was living in a matchbox apartment on a military base.

But I had her.

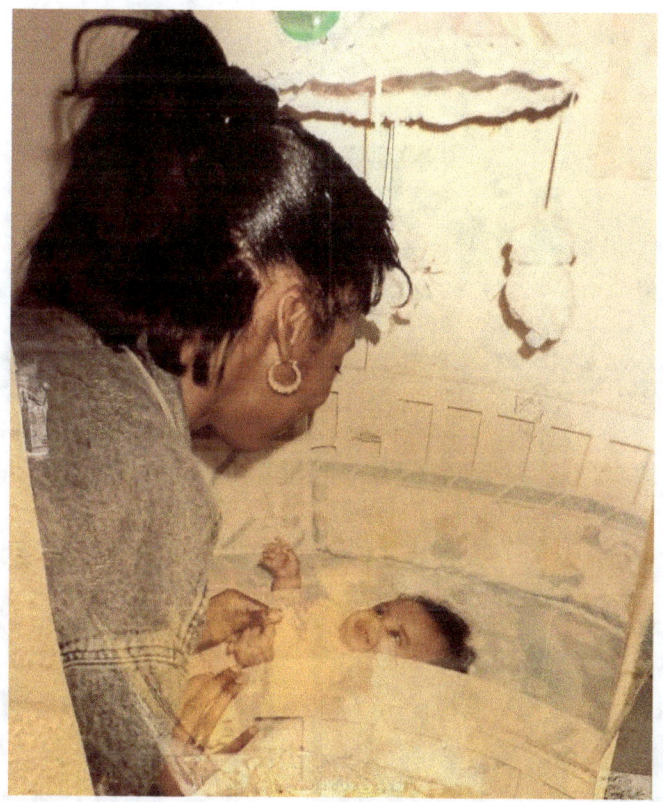

And she was everything. I would spend evenings with her on a soft blanket, singing, talking, watching her discover the world. She would grab my finger and stare at my face like I was the most fascinating thing she had ever seen.

Before bed, I would hold her close and whisper prayers. "God, thank You for keeping her safe. Thank You for bringing her into my life. Please protect her. Please give her everything I never had." Then I would kiss her forehead and think, this is why I'm here. This is why I survived. For her.

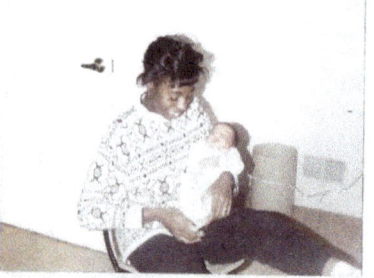

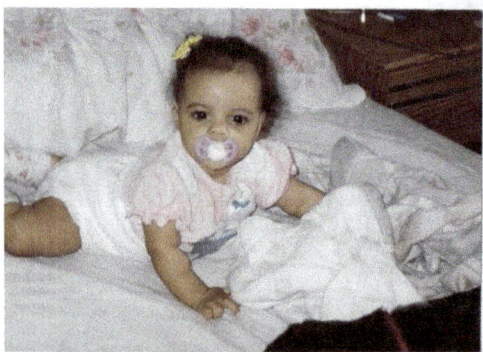

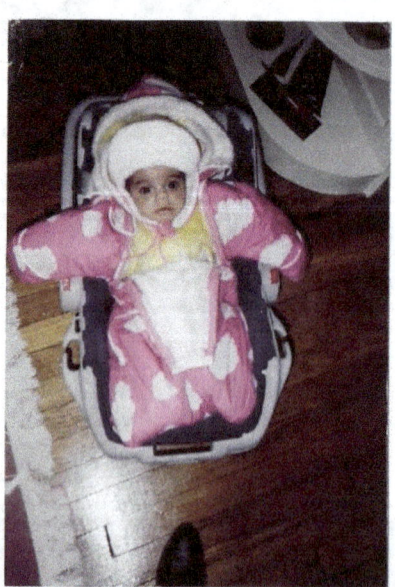

Chapter 29: The SCIF to Nowhere

February 1992. Fort Meade, Maryland.

I finally got my Top-Secret clearance. After months of waiting, months of background checks and interviews and paperwork, it was finally adjudicated. Which meant I could start my actual job. Working at the NSA.

I had been doing administrative work up until that point. Filing. Typing. Running errands. Boring, mindless tasks that anyone could do. But now that I had my clearance, I was supposed to be doing intelligence work. The job I had trained for. The reason I had joined the Army in the first place.

Working at NSA felt like stepping into a world I had only seen from a distance—the mirrored black towers rising out of Fort Meade like giant puzzle pieces, glinting in the morning sun. From above, the place looked like a fortress of reflections—no windows, no cracks, just sleek panels of secrecy surrounded by miles of cars lined up like tiny soldiers waiting for orders. Inside, everything gleamed: polished floors, bustling hallways, and the sound of heels clicking on tile like code.

I would picture myself walking those halls, attaché case in hand, moving like I was headed straight for world domination—part Maxwell Smart, part Jason Bourne, part U.N. ambassador—important enough to carry secrets in a locked case. I did not even know what went inside one of those cases, but I wanted one just the same. It was not vanity; it was vision. In my mind, that case held my future—a woman with purpose, confidence, and clearance.

For a moment, sitting there in that black-mirrored government building, I almost believed I had made it. Everything I had fought through—the training, the ELINT (electronic intelligence) school, the tactical field training in Germany—was leading me here. I did not know what was waiting around the corner yet. I was just a young

Black woman in uniform, still believing that if I worked hard enough, I could become her—the woman with the attaché case and a mission. Except it did not work out that way.

ॐ

On my first day with clearance, they took me to the Sensitive Compartmented Information Facility (SCIF). A secure room where classified work happened. I had heard about SCIFs in training—high-security spaces, monitored and controlled, where the real intelligence work was done. I had imagined something impressive. High-tech. Important. But when they opened the door and showed me where I would be working, I almost laughed.

It was a closet. Not literally. But close. A tiny, windowless room barely big enough for two desks, two chairs, and two soldiers. Me and one other guy. A white kid, maybe twenty years old, who looked about as thrilled to be there as I was.

"This is it?" I asked. "This is it," the sergeant who'd escorted me said. "What are we supposed to be doing?" He shrugged. "Whatever they tell you to do." "And what's that?" "I don't know. I don't work in here." And he left.

I sat down at one of the desks. The other soldier—I don't even remember his name—sat at the other. We looked at each other. "So," I said. "What do we do?" "I don't know," he said. "I've been here for two weeks, and nobody's told me anything." "Nothing?" "Nothing."

We sat there. In that tiny, windowless room. With nothing to do. No assignments. No tasks. No orders. Just... nothing. Hours passed. I read a manual someone had left on the desk. He fell asleep. Nobody came to check on us. Nobody gave us work. We were just... there.

It was the most ridiculous thing I had ever experienced. After everything I had been through—basic training, AIT, Germany,

Desert Storm—I was sitting in a closet doing absolutely nothing. This was the intelligence work I had signed up for?

After a few days of this, I started bringing correspondence courses with me. The Army offered them—self-paced courses you could take to earn promotion points or college credit. You'd get a packet in the mail with a textbook and assignments. You could work through them on your own time, then mail back the completed tests. It was something to do. Something to keep my brain from rotting in that windowless box. So, I studied. And I waited for someone to tell me what my actual job was supposed to be.

But nobody ever did. Weeks went by. Then months. And I just kept showing up to that SCIF, sitting at that desk, doing correspondence courses, and wondering what the hell I was doing with my life.

One day, I decided I needed to do something bigger. Something that would actually change my situation. So, I started researching the Green to Gold program.

Green to Gold was an Army program that allowed enlisted soldiers to go to college full-time, get their degree, and become commissioned officers. The Army would pay for your tuition, give you a stipend and let you go to school like a normal college student. And at the end, you'd graduate as a second lieutenant. It was exactly what I needed. A way to get the college education I had always wanted. A way to become an officer instead of staying enlisted. A way to actually use my brain instead of wasting away in a SCIF closet.

I put together my packet. It was not easy. But I had everything I needed. You had to have recommendations, a clean record, proof of academic potential and, all kinds of other documentation. I had served in Desert Storm. I had awards. Badges. A Southwest Asia Service Medal. A Kuwait Liberation Medal. I had been through hell and survived. If anyone deserved a shot at Green to Gold, it was me. I submitted the packet to my company commander. A white man. Captain somebody. I don't even remember his name. And I waited.

Then came the summons. A few weeks later, he called me into his office. "Specialist Adams," (I had received a promotion) he said, not looking up from his desk. "Yes, sir." "I reviewed your Green to Gold packet." My heart started pounding. "And..."

"I'm not going to recommend you." The room tilted. "What?" "I'm not recommending you for the program." "Can I ask why, sir?" He finally looked up. "You haven't been to enough boards."

I stared at him. "Boards, sir?" "Yes. Promotion boards. Soldier of the month boards. That kind of thing. You need more leadership experience." "Sir, I was in Desert Storm. I served in a combat zone. I—" "I'm aware of your service, Specialist. But that doesn't change the fact that you haven't demonstrated enough leadership potential through formal boards."

I could not believe what I was hearing. I had been to war. I had survived things most of these officers would never have to face. I had crawled through sandstorms, dodged Scuds, and watched death brush past at the tank battle of 73 Easting—and somehow, that's not enough leadership? A white male soldier, who was in my unit, who'd never even set foot in Desert Shield or Desert Storm got his Green to Gold request approved—fast-tracked, no questions asked. A soldier who could not even perform under normal conditions because he was severely allergic to everything, including the grass. But me? After everything I survived, I was told I did not "demonstrate sufficient leadership potential."

I remember thinking: *What the hell does survival count for, then?* And he was telling me I was not qualified because I hadn't competed in some bullshit dog-and-pony show boards?

"Sir," I said, trying to keep my voice steady. "With all due respect, I think my combat experience should count for something." "It does count for something," he said. "But it's not enough. You need to round out your record with other experiences. Come back in a year or two after you've done some boards, and we'll talk."

A year or two. I did not have a year or two. I had a baby. I was barely surviving on my enlisted salary. I needed to move forward, not wait around hoping someone would eventually decide I was worthy.

"Is there anything else, Specialist?" "No, sir." "Dismissed." I stood, saluted, and walked out of his office. And I wanted to scream.

ॐ

This was it. This was the moment I realized the Army would never give me what I needed.

The truth hit hard after that meeting—I was never going to be on the right side of luck in the Army. Almost shipped to Wild Chicken. Sent to Desert Storm. Survived 73 Easting and came home just to be told I was not enough.

It felt like there was a bullseye stamped across my file in big red letters: *Black Girl. Make life a living hell for her. Send her to the worst duty assignments you've got. And watch her keep standing anyway.* A Black woman. A single mother. Someone who did not fit the mold.

I could do everything right—serve in combat, earn my medals, keep my head down—and it still wouldn't be enough. Because the system was not designed for people like me. It was designed to keep people like me in our place.

That night, I went home to my tiny apartment. I picked up Deidra. Held her close. Breathed in her baby smell. And I thought: *I can't do this anymore.*

I can't keep fighting a system that doesn't want me. I can't keep sacrificing my life, my health, my sanity for people who don't care.

And I started thinking about leaving.

Chapter 30: Orders To Turkey and Her Final Choice

Spring 1992. The orders came in April. I had been at Fort Meade for about seven months. Deidra was five months old. And I thought, naively, that maybe things were finally settling down. That maybe I could just do my job, raise my daughter, and figure out my life one day at a time. But the Army had other plans.

I was called into my company commander's office. Again. The same commander who'd rejected my Green to Gold application. "Specialist Adams," he said, looking at a file on his desk. "Your next assignment came through." My stomach dropped. "Sir?" "Sinop, Turkey. One-year unaccompanied tour. You'll deploy in August."

The room went silent.

Unaccompanied. That meant I could not bring Deidra. That meant I would be gone for an entire year. That meant I would have to leave my five-month-old baby with someone else. For a year.

"Sir," I said carefully. "I have a dependent. A baby. She's five months old." "I'm aware, Specialist. That's why you'll need to complete a Family Care Plan." "A Family Care Plan?" "Yes. It's a document outlining who will care for your dependent while you're deployed. You'll need to designate a guardian, provide their contact information, and get it notarized." He slid a packet of paperwork across the desk. "This needs to be completed and submitted within thirty days. So I suggest you get started."

I took the paperwork. Walked out of his office. And I felt like I was going to fall out.

A Family Care Plan. Who was I supposed to designate? My mom? She had just given birth to Lucky, and we were supposed to be raising our babies together. This threw a whole monkey wrench in our plans. On top of that, she did not have the capacity to care for two babies for a year. My aunt Teresa? She had her own life, her own family. She had already helped me so much. I could not ask her to take on a year-long commitment. Dana? She was Deidra's babysitter, not her guardian. And besides, she had her own family to take care of.

I sat in my car in the parking lot, staring at the paperwork, and I could not stop thinking about what this would mean. Leaving Deidra for a year. Missing her first steps. Her first words. Her first everything. Trusting someone else to raise her for twelve months while I was halfway around the world. And for what? For a career that had given me nothing but pain? For an Army that did not value me, did not protect me, did not care about me?

And then I saw it. In my mind. Clear as day. I saw my future self. The woman who left and went to Turkey. The one who left Deidra behind. And she was broken. Destroyed. Hollowed out by guilt and grief and the knowledge that she had chosen the Army over her child. I saw her coming home to a daughter who did not recognize her. Who cried when she tried to hold her. Who'd bonded with someone else because her mama was not there. I saw the damage. The trauma. The irreparable harm.

And I thought: *No.*

Not me.

That will not be my story.

I had made the choice to say no once before. In the back of that Humvee with Strychland, when I saw my future and refused to let him destroy me. And I had survived. This was the same thing. I was seeing my future. And I was choosing to refuse it.

❧

I went home that night and held Deidra. Looked into her perfect little face. Watched her smile at me with that gummy, toothless grin. And I said out loud: "I'm not leaving you."

The next day, I started researching my options. And I found it. A way out. Hardship discharge. The Army allowed soldiers to request separation under certain circumstances, family hardship being one of them. If you were a single parent and could not complete a Family Care Plan, you could apply for a hardship discharge. It was not guaranteed. But it was possible.

I filled out the paperwork. Explained my situation. Single mother. No support system capable of caring for my infant daughter for a year. Unable to comply with the requirement for an unaccompanied tour. I submitted it. And I waited.

It took less than a month. The approval came through in early June. Honorable discharge. Effective July 14, 1992. I was getting out.

I should've felt relieved. And part of me did. But another part of me felt... scared. Because I had just walked away from the only plan I had ever had. The Army was supposed to pay for my college. Supposed to give me a career. Supposed to be my ticket to a better life. And now I was leaving with nothing. No degree. No job prospects. No plan. Just me and Deidra. And I was moving on instinct alone for what came next.

❧

My last day in the Army was July 14, 1992. I out-processed. Turned in my gear. Signed the final paperwork. And then I got in my car to drive off base. And the car broke down. Right there in the parking lot. Engine dead. Won't start.

I sat there, staring at the dashboard, and I almost laughed. Of course. Of course this would happen. On my last day. As if the universe was saying: *You sure about this?*

I got the car towed and fixed–eventually. But the symbolism was not lost on me. Leaving the Army felt like driving into the unknown with a broken-down car. No map. No GPS. No guarantee I would make it. Just faith. And hope. And the belief that I had made the right choice.

That night, I lay in bed with Deidra beside me. She was asleep, her tiny chest rising and falling. And I whispered to her: "I chose you. Over everything else, I chose you. And I'm going to make sure you have the best life I can give you. I promise."

I did not know how I was going to keep that promise. Did not know how I was going to pay bills or go to college or build a career. But I realized one thing for sure: I had survived everything the Army had thrown at me. Basic training. Germany. Desert Storm. Assault. Retaliation. Isolation. I had survived it all. And I would survive this too because I had Deidra. And she was worth fighting for.

Almost three years. That's how long I had been in the Army. August 1989 to July 1992. Two years and eleven months. But I had been to combat. I had been to war. That makes every month feel like a year. Two years and eleven months that shaped me, broke me, and rebuilt me. Two years and eleven months that gave me the greatest gift I ever received.

And now it was over. And I was free.

ॐ

I had witnessed a blanket party back at Fort Devens during AIT—the girl who refused to shower, dragged to the latrine and forcibly washed. At the time, I understood it. Even agreed with it. Someone was making life harder for everyone else. Someone needed to be corrected. So, they corrected her. It never occurred to me then that I was about to spend the next three years in my own blanket party.

Not the literal kind. No one threw a blanket over my head. No one dragged me anywhere. But the punishment was just as real. Just as coordinated. Just as designed to break me down for not conforming.

For being Black in a white Army. For being a woman in a man's world. For getting pregnant when I was supposed to stay mission-ready. For speaking up when I was supposed to stay quiet. For saying no when I was supposed to roll over and open my legs. For surviving when they expected me to break.

The entire military became one long blanket party. And I was the one getting beaten. Staff Sergeant Strychland beat me down with retaliation, water details, tripod duty and sexual harassment. Staff Sergeant Bleach-Bottle Blond beat me down with duty rosters that always started at A. The EO officer beat me down by telling me it was my word against Strychland's. The company commander beat me down by rejecting my Green to Gold application. The system beat me down by giving me orders to Turkey and telling me to leave my five-month-old baby for a year.

They all participated. Some actively. Some passively. Some because they hated me. Some because they did not care enough to stop it. But they all threw their punches. And they all expected me to just take it. To conform. To comply. To break.

But here's the thing about blanket parties: They only work if the person being beaten stays down. If they give up. If they decide it's not worth fighting anymore.

And I did not stay down. I got back up. Every single time.

When Strychland tried to assault me, I said no. When Staff Sergeant Bleach-Bottle Blond tried to break me with endless duties, I kept going. When the EO officer told me to forget it, I survived anyway. When the commander rejected my application, I found another way out. When the Army tried to separate me from my daughter, I walked away.

They wanted me to break. To quit. To prove I did not belong. To validate their belief that Black women could not hack it. But I did not give them that satisfaction. I survived. I left on my own terms. I chose my daughter over their system. And I carried the truth with me, locked away for thirty plus years, yes, but never forgotten.

Here I am. Still standing. Still refusing to be erased.

୬

To my daughter: *I've always told you that you were planned and conceived in love. And although it was ill-planned by a teenager with little life experience, I was in love in that moment. The main reason I prayed over your life—asking God to give you above and beyond what two earthly parents could provide—was so you'd never have to choose between a rock and a hard place the way I did. I did not want you to have to join the military just to escape poverty and get an education. I wanted you to have real choices, not just survival options.*

I think about that nineteen-year-old girl a lot now. The one who made a mistake and owned it completely. No deflection. No excuses. Just raw honesty, even when it cost her everything. The one who endured abuse, retaliation, isolation—and never once played the victim. She had no advocate. Not one person to stand up for her, to say "this isn't right," to tell her she mattered. Just her and God.

She handled herself with dignity when no one gave her any. She kept going when everyone wanted her to break. She made the impossible choice—choosing you—and she did it alone.

If I could go back, I would stand right beside her and tell her everything she needed to hear:

You're going to be okay. Better than okay. You think you're alone, but you're not. God sees you. He's been there the whole time. And He's preparing you for something bigger than survival. The way you're handling this? The dignity you're keeping when they're stripping yours away? That's not weakness. That's warrior strength. You're going to be a mother. A good one. The kind who fights for her daughter the way no one fought for you. You're going to build a life. On your own terms. Not their terms. You're going to tell this story one day, and people are going to see you the way I see you right now—Brave. Dignified. Unbreakable. You did everything right, even when everything went wrong. And I'm so proud of you.

I can't be the advocate she needed then. But I can give her this: Recognition. Compassion. The truth that she handled it with more grace than anyone had the right to expect. That's what this memoir is, too. Not just defiance. Not just testimony. It's finally giving that girl what she deserved all along—someone to say: *I see you, Darnnell. You did good. And your story matters.*

You asked what the war was like.

It was fraught with terror, alienation, annihilation, and isolation.

But God! In the sandstorm, in the duties, and in the foxhole, He covered us both with His mighty pinions so the war could not claim us. He gave me mercy in all those unexpected acts of random kindness. And then He brought me safely home so that I could bring you safely into this world.

You are the victory.

Edna, Deidra, Lucky, Aunt Teresa

Edna, Lucky and Deidra in Disney World

Darnnell, Lucky and Deidra in Disney World

Aunt Teresa, Edna and Deidra

Darnnell and her sister Lucky and brother Isaiah and daughter Deidra

Darnnell & Deidra

Edna Lucky Darnnell Deidra

Lucky

Charles + Deidra

Terrence + Darnnell

Deidra at her college graduation

Savannah College of Art and Design

by authority of the Board of Trustees
and on the recommendation of the faculty
hereby confers upon

Deidra Darnnell Adams

the degree of

Bachelor of Fine Arts

upon satisfactory completion of the course of study in

Fashion

In evidence whereof this diploma is awarded and attested
by the seal of the College and the signatures authorized by the Trustees.

Savannah, Georgia   May 28, 2015

Chair, Board of Trustees   President of the College

You did it Deidra!!! I'm so
proud of you, Love Mom

Strayer University

By the virtue of authority of the Board of Trustees and the
recommendation of the Faculty
has conferred upon

Darnnell Adams

the Degree of

Bachelor of Science
Database Technology

with all the rights, honors, and privileges thereto pertaining.
In witness whereof, this degree is granted bearing the seal of
Strayer University, Washington, D.C.

Given this twenty-sixth day of March in the year two thousand and seven.

Chairwoman, Board of Trustees   University President

Summa Cum Laude

You did it Darnnell!!! I'm so proud of you,
Love Dee

Sister Lucky, brother Isaiah, daughter Deidra, brother Ricky

Deidra's sketch of herself
at Dana's house

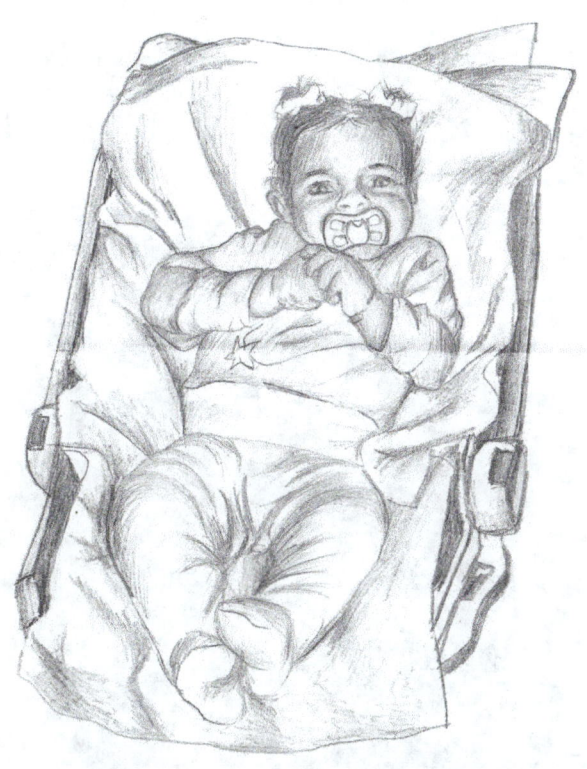

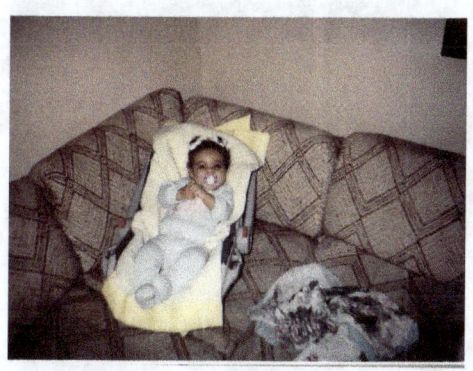

Deidra's sketch of her great grandmother Josephine

Deidra's sketch of her grandma Cheryl

Uncle Kenny (RIP)

Deidra's sketch of Uncle Kenny

Lucky and Deidra
after HS graduation

Lucky and Deidra on
prom night

Terrence and Deidra

Epilogue: A Daughter's Voice

I did not know what I would find when I finally sat down to read my mother's story from beginning to end — not the fragments I grew up with, not the folded photographs or the half-told memories, but the full, unfiltered truth. What I discovered was both more harrowing and more awe-inspiring than anything I could have imagined.

Some stories don't get easier with time. They just get heavier. And the decision to finally set them down, to let someone else hold them for a while, is an act of trust and courage I'm still learning to comprehend.

༒

Nothing prepared me for the moment she described her near-death experience in Kuwait — the chaos of a tank battle erupting without warning, the frantic orders to abandon their Humvees, the desperate scramble to dig foxholes in the dark. She survived the sixth most massive tank battle in U.S. history — the Battle of 73 Easting — without even knowing the significance of what she had been thrust into. She was barely 115 pounds, without the physical strength the Army prized, without support from leadership who should have been watching over her — and still, she lived.

When she wrote about emerging from that sand-filled grave, stepping onto a road lined with decimated tanks and burned bodies, I felt a heaviness settle in my chest. I never knew she had carried those images for thirty-four years. I did not know she had seen the faces of young enemy soldiers — boys, really — kneeling, starving, surrendered. She did not see enemies that day. She saw victims. And I understand now why the nightmares never fully left her. Why she takes medication. Why loud noises still startle her body into remembering. They followed her home. Even now, she still dreams of being alone in a war zone, of bombs falling and, of searching for shelter that didn't exist. But silence has a cost. And my mother paid it in her body, in her spirit, in the ways she learned to make herself small.

The hardest part, though, was not the war. It was what happened within her own ranks.

଼ୱ

It broke something in me to read about the night her sergeant tried to rape her in the back of a Humvee, how quickly the isolation descended, how rumors spread faster than protection. It stunned me to learn that both the man who would become my father and the man she had once loved were right there — traveling with her team — and yet neither noticed how her spirit shifted after that night. They were lost in their own hurt, their own assumptions. But she was alone.

And still — she fought back. She said "No" twice. She made a choice in that moment not to carry that violation into the rest of her life. I did not know that kind of strength lived in her body so early, so fiercely. Reading that felt like discovering a new definition of the word mother.

So much of her story is about endurance that went unseen: Being ostracized by leaders who should have supported her. Being punished for lies spread by the man who harmed her.

Performing grueling labor while trying to hide a pregnancy that could have cost her everything. Surviving a war abroad only to fight a quieter, more exhausting one at home — with the VA, with unexplained illnesses, with decades of bureaucratic dismissal. And yet she raised me with love, with intention, with more care than she ever received. There were moments while reading when I thought: *Anyone else would have broken under all of this. Anyone else might not have survived.*

But she did.

2✍

I understand her now in a way I never did before — as a soldier, as a woman, and as a mother whose resilience came at a cost she did not always let me see. I see why she fought so fiercely for her benefits, why she clung to faith, why she rebuilt herself more than once. I see how trauma shaped her, and how she refused to let it define the rest of her life.

It was not until recently that she started doing the real work of healing. Not the kind that comes from time or distance, but the kind that requires you to turn around and face what tried to destroy you. She found her way to a network of Black women counselors, even some at the VA who understood what she had been through without her having to explain it. For the first time in her life, my mother has a community of Black women who see her, not as a liability or a problem, but as a survivor. A veteran. A woman who deserves care.

2✍

What I inherited from her is not just resilience — but faith. The quiet, stubborn kind that pulls you back to God after every devastation. The kind that rebuilds you when the world has tried to break you into pieces. As a teenager, I did not understand her return to faith. Now, in my thirties, I do. The Holy Spirit meets you when you are finally ready to listen. And in reading her story, I realized I am on that path too.

My hope for this book is simple but profound: I want Black women — especially those who served — to feel seen. To know they are not forgotten or invisible. To know that their pain is real, their stories are worthy, and their silence does not have to be permanent.

If even one woman feels empowered to speak her truth after reading my mother's, then this book has done something sacred. Because systems only maintain their power when people stay quiet. And Black women have been quiet for far too long.

As for how I want readers to leave these pages — maybe the truth is that I don't want to tell them how to feel. I only want them to understand what I now understand:

That my mother survived a war abroad, a war within her own unit, and a war with the institutions that should have protected her — and she still found the strength to create a life, to raise a daughter, and to reclaim her voice.

৯৬

What I understand now is that survival doesn't always look heroic. Sometimes it looks quiet, tired, persistent, and unseen. My mother survived in all those ways — and she kept going.

I can't rewrite her past.

But I can honor it.

And I will.

Deidra Wilson
2025

ABOUT THE AUTHORS

At seventeen, Darnnell Reese dreamed of becoming the "smart girl carrying the attaché case"—a woman in a sleek suit attending high-level meetings with classified documents. When an Army recruiter promised her a career in military intelligence, she believed that dream was within reach.

Darnnell served as a Military Intelligence Analyst (MOS 98J) with the 2nd Armored Cavalry Regiment during Operation Desert Storm, surviving the Battle of 73 Easting—the largest tank battle since World War II. She is one of the few Military Intelligence soldiers authorized to wear the 2nd ACR combat patch, a distinction earned under fire in the Iraqi desert.

But she earned it the hard way. The Army gave her a blanket party during Desert Storm that left her beaten and broken, a SCIF closet with nothing to do, a captain who denied her Green to Gold application because she hadn't "been to enough boards," and orders to a one-year unaccompanied tour in Turkey that would have separated her from her infant daughter. What the Army could not do was extinguish her indomitable will to achieve her sacred goals and dreams, no matter how lofty they seemed.

After leaving the military, Darnnell did not just rebuild her life—she exceeded every expectation the Army ever had for her. She earned her Bachelor of Science degree in Database Technology from Strayer University, graduating *Summa Cum Laude* with a 4.0 GPA and President's List honors every quarter. She mastered UNIX shell scripting and database administration, transforming herself into the technical expert she had always envisioned.

Her federal career spanned more than two decades of distinguished service. At the Internal Revenue Service, she managed a 20,000-user procurement database and wrote automated scripts that reduced hours of manual work to minutes. At the Treasury Inspector General for Tax Administration (TIGTA), she served as a CyberSecurity and Fraud Forensic

Investigative Analyst, using tools like EnCase and Intella to investigate federal employee misconduct—work that was cited in a report to Congress. She held a Top-Secret security clearance and earned consistent "Exceeded Expectations" performance ratings.

Darnnell's career culminated at INTERPOL Washington, where she served as a Supervisory Investigative Analyst leading a team of twelve analysts, contractors, and law enforcement detailees in highly complex international investigations. She became a certified instructor for the Foundations of Intelligence Analysis Training (FIAT) program, teaching the next generation of law enforcement intelligence professionals.

The girl who dreamed of carrying the attaché case finally got it—and filled it with credentials that captain could never have imagined for her.

Blanket Party in Desert Storm is the third book in her "Victorious With God" series, following *Victorious! Defeating Bullies and Giants God's Way* and *In All Seriousness... Totally Funny Bible Stories*. This memoir is co-authored with her daughter, Deidra Wilson, a Savannah College of Art and Design (SCAD)-trained fashion designer who contributed the prologue and epilogue, offering a daughter's perspective on the legacy of military service.

Darnnell is now happily retired and lives in Fort Washington, Maryland, with her husband Terrence, a federal police officer. She remains committed to validating the experiences of veterans whose stories have been overlooked, and to testifying about the divine protection that carried her through every storm.

Deidra Wilson is a fashion designer and graduate of Duke Ellington School of the Arts in Washington, D.C. and the Savannah College of Art and Design (SCAD) with over ten years of experience in the fashion industry. Deidra was blessed with extraordinary visual gifts—she can draw a person's likeness freehand with impeccable detail, a talent her mother never possessed and can only marvel at as divinely given. Her artistic abilities may also be a result of her mother's love for Pictionary. She has designed standout collections for Joe's Jeans, Good American, Favorite Daughter, Athleta, JustFab, ShoeDazzle, and DVF.

As the daughter of a Desert Storm combat veteran, she grew up witnessing her mother's fight for recognition and healing, never fully understanding the cost until she asked her mother to share her story. That conversation became the catalyst for this memoir. But Deidra's greatest contribution was not her artistry—it was her hunger to know. This book tells the story of her genesis, and she refused to let any detail go uncaptured. She was instrumental in shaping how this story evolved, pressing her mother for every remembrance and asking the questions only a daughter seeking to understand her own origins would think to ask. She authored the prologue and epilogue, ensuring that the full truth of how she came to be would finally be told.

Deidra is passionate about amplifying the voices of Black women veterans whose service has been erased from history. She lives in Los Angeles with her husband Charles, Founder and CEO of Ghost Dog Productions and Co-Producer of Warner Bros.' *All American.* His love, kindness, unwavering support, and fierce protectiveness shield her from the harsh realities that Black women navigate daily, creating the safe space that made it possible for her to help bring her mother's story to the world.

RESOURCES FOR VETERANS

If You Need Support

If you are a veteran experiencing Military Sexual Trauma, Gulf War Illness, or need support, please know that help is available:

VA Military Sexual Trauma (MST) Support: Every VA medical center has an MST Coordinator. Call 1-877-222-8387.

Veterans Crisis Line: Call 988 (then press 1), text 838255, or chat at VeteransCrisisLine.net.

Gulf War Veterans Information: Visit www.publichealth.va.gov/exposures/gulfwar.

Women Veterans Call Center: 1-855-829-6636 (1-855-VA-WOMEN).

You served. You survived. You deserve care.

If you're a veteran struggling to get the care you deserve, I'm happy to share what worked for me—reach out at Darnnells@gmail.com.

OTHER BOOKS BY DARNNELL REESE

VICTORIOUS! Defeating Bullies and Giants God's Way

(Victorious With God Series, Book One)

In All Seriousness...Totally Funny Bible Stories

(Victorious With God Series, Book Two)

Available on Amazon and wherever books are sold.

Thank you for reading our story. If it moved you, we'd be grateful if you'd leave a review on Amazon or Goodreads. Your words help other readers — and other veterans — find this book. For Kirkus Reviews, book news, and more from the Victorious With God series, visit: https://reeseauthor.com/ Kirkus page: https://www.kirkusreviews.com/author/darnnell-reese-1/

Amazon Author Page: https://amazon.com/author/darnnellreese Kirkus page: https://www.kirkusreviews.com/author/darnnell-reese-1/ and Amazon Author Page: https://amazon.com/author/darnnellreese

www.ingramcontent.com/pod-product-compliance
Lightning Source LLC
Chambersburg PA
CBHW070912130626
46555CB00001B/104